SCHEDULING

for Home Builders

WITH MICROSOFT PROJECT

SCHEDULING
for Home Builders
WITH MICROSOFT PROJECT

David Marchman and Tulio Sulbaran, PhD

A Service of
NAHB
NATIONAL ASSOCIATION
OF HOME BUILDERS

Scheduling with Microsoft® Project

BuilderBooks, a Service of the National Association of Home Builders

Courtenay S. Brown	Director of Book Publishing
Torrie Singletary	Production Editor
MacDesigns, Inc.	Cover Design
Pine Tree	Composition
McNaughton & Gunn, Inc.	Printing

Gerald M. Howard	NAHB Executive Vice President and CEO
Mark Pursell	NAHB Senior Staff Vice President, Marketing & Sales Group
Lakisha Campbell	NAHB Staff Vice President, Publications & Affinity Programs

Disclaimer
This publication provides accurate information on the subject matter covered. The publisher is selling it with the understanding that the publisher is not providing legal, accounting, or other professional service. If you need legal advice or other expert assistance, you should obtain the services of a qualified professional experienced in the subject matter involved. Reference herein to any specific commercial products, process, or service by trade name, trademark, manufacturer, or otherwise, does not necessarily constitute or imply its endorsement, recommendation, or favored status by the National Association of Home Builders. The views and opinions of the author expressed in this publication do not necessarily state or reflect those of the National Association of Home Builders, and they shall not be used to advertise or endorse a product.

Printed in the United States of America

10   09   08   07   2   3   4   5

ISBN-13: 978-0-86718-621-5

Cataloging-in-Publication Information

For further information, please contact:

Library of Congress Cataloging-in-Publication Data

Marchman, David A.
  Scheduling with Microsoft project / David A. Marchman, Tulio Sulbaran.
    p. cm.
  ISBN-13: 978-0-86718-621-5

  1. House construction—Data processing. 2. Production scheduling—Data processing. 3. Microsoft Project. I. Sulbaran, Tulio A. II. Title.

  TH4812.M2364 2006
  690'.8370685—dc22

                                            2006047559

National Association of Home Builders
1201 15th Street, NW
Washington, DC 20005-2800
800-223-2665
Visit us online at www.BuilderBooks.com.

Contents

Figure List

Chapter 5

Chapter 6

About the Authors

David A. Marchman is full professor and associate director in the School of Construction at the University of Southern Mississippi. His present teaching responsibilities include construction organization and the senior capstone course. He holds a degree in Building Construction from the University of Florida and has extensive construction experience (residential, commercial, and industrial). He has published several other scheduling books and has taught scheduling seminars for residential builders over many years.

Tulio A. Sulbaran, PhD, teaches estimating, scheduling, project management, and other construction courses at the School of Construction at the University of Southern Mississippi. He holds a doctorate degree from Georgia Institute of Technology and has several years of international work experience in the architecture, engineering, and construction. Dr. Sulbaran serves as a consultant for numerous companies, including URS Corporation, W.G. Yates Construction, Superior Asphalt, and Mississippi Power, among others. He has been very active in organizations such as the Associated School of Construction, the American Council for Construction Education, the American Society for Engineering Education, and the Associated Builders and Contractors. Additionally, he has taught construction-related seminars and workshops at nationally and internationally recognized conferences.

Acknowledgments

David A. Marchman thanks his wife, Janet S. Nelson; his son Dane and his family (Holly, Forrest, Sawyer, and Marilea Marchman); and his daughter Dee and her family (Josh, Kalah, and Kalel Turnage) for their support. Professor Marchman also thanks Dr. Sulbaran for his collaboration on this project and Professor Desmond Fletcher for his support.

Tulio A. Sulbaran expresses gratitude to his undergraduate students and the important role his student assistants, Marc Brana, Scott Corage, Justin Nosser, and Arlys Silva, played in helping him focus on this book. He also thanks the many construction companies and construction organizations that provided priceless information for this book as well as opportunities for personal and professional growth. Dr. Sulbaran gives a special thanks to Professor Marchman for the opportunity to participate in preparing this book. Dr. Sulbaran also thanks his mother Alida Gonzalez, his wife Virginia, his son Tulio Nicolas, and his daughter Virginia Valentina for their nurturing, support, and understanding.

Introduction

All home building projects entail coordinating the project plans and specifications, construction materials, craftspeople, construction equipment, and trade contractors to build the home. How you utilize these resources determines whether your project will run efficiently and progress in an organized fashion or end up in total disarray. By developing a schedule at the beginning of the project, you can better plan and arrange the tasks necessary to complete the project. An organized schedule is a builder's best tool for controlling the project parameters and costs of building a home.

Microsoft® Project

Microsoft® Project allows you to define tasks and view the interrelationships between them to develop a home construction project schedule. You can also use the software to define resources and their limits and costs to determine the cash flow for the project. A key feature of the software is its ability to sort the schedule by alternate sort criteria. For example, you can perform a sort by task responsibility that shows the craft foreperson or trade contractor who is responsible for the task completion.

Another key feature of Microsoft® Project is its flexible, easy to use hard-copy print capability. Effective scheduling involves disseminating information and receiving feedback from all key parties involved in the project. If you can't clearly communicate your schedule to key parties involved in the project, the schedule is useless.

Using the CD

The CD provided with this book includes sample schedules that you can use to practice the skills that are presented in each chapter. You must have Microsoft® Project 2003 installed on your computer to access the files on the CD. There are two versions of Microsoft® Project—standard and professional. The difference between the versions, besides price, is that the standard version is designed for a single user, whereas the professional version is designed to be used on a server and accessed by multiple users. The practice exercises in this book were created using Microsoft® Office Project Standard 2003. You can download a 60-day trial version of the standard software online at http://www.microsoft.com/office/project/prodinfo/trial.mspx.

Microsoft® Project is a registered trademark of the Microsoft Corporation. Screen shots within the book are reprinted with permission of Microsoft.

CHAPTER 1

Introduction to Scheduling

Objectives

Home building projects are becoming more complex and more costly. Therefore, you must pay greater attention to how you manage your time and resources (specialized labor, materials, and construction equipment) over the duration of the project. Planning and controlling resources and pacing construction allows you to meet your contractual deadlines and ensure a quality product. This chapter will help you understand the fundamentals of an effective home building schedule by explaining the following Microsoft® Project elements:

- tasks
- phases
- milestones
- Gantt Charts and reports
- rough logic diagram
- schedule

Components of an Effective Schedule

The old adage "time is money" is particularly true in home building. The sooner a home is completed, the sooner you can move on to the next project. Because your profit motive drives the project, another old adage applies: "Time is of the essence." Delays can be costly in terms of time and lost interest on construction loans. Ultimately the work will be accomplished, but proper scheduling allows you to better allocate your resources and save money.

You are selling your firm's management ability to put the buyer's project in place. If you can save time, you can increase your volume of work and your profitability. The efficient construction company can produce more work per year with the same management staff. Time *is* money!

Estimates and Schedules

The four steps to successful residential project control are planning, scheduling, monitoring, and controlling the project. Effective project management means your project is "under control" rather than "out of control." All projects have two primary control documents: the estimate and the schedule. The *estimate* defines the scope of the project in terms of quantities of materials, work hours of labor, and hours of equipment use. Since all these resources can be measured in dollars, the estimate also serves as the cost budget.

The *schedule* controls how you spend your time. Since time is money, the resources and the time frame over which they are expended are directly related, and the elements of the estimate and the schedule are interrelated. The information from one document has implications for the other. The estimate computes resources needed over the specified time and the schedule defines the time needed given the specified resource. Successful home builders use these documents to conceptualize the project before any actual construction begins. This is the essence of planning.

Estimating, and to an even greater extent, scheduling, allow you to anticipate surprises and solve problems during the planning process (before resources are expended), rather than on the jobsite during the construction phase. Proper planning is necessary to make the construction process run smoothly, minimize surprises, and prevent expending costly resources with poor results.

Communication

Only the smallest home building project is completed by a single individual. Most projects require the services of many different experts and various contractual relationships. Therefore, you need to ensure effective communication at every stage in the process. In order for a project to run smoothly all parties need to be on the same page at the same time. This requires constant communication so that all participants know the plan, understand their responsibilities, and are aware of your requirements and expectations. Keeping everyone informed is one of the critical functions of the scheduling process.

Tasks

One of the first steps in developing a schedule is to identify the tasks that must be completed. A *task* is a concrete step that's required to meet a project goal. Effective scheduling requires more detailed preparation as a particular activity gets closer to actual installation.

Tasks Defined

Residential construction projects are comprised of a number of individual tasks that must be accomplished to complete the project. Tasks (other than milestones) have five specific characteristics:

- time consuming
- resource consuming
- have a definable start and finish
- assignable
- measurable

Task are defined by the project estimate, available historical information, and the project team's experience.

Task Names

The *task name* describes the activity. Task names must be clear, concise, and universal (they must mean the same thing to all parties using the schedule). This includes your workforce, trade contractors, the owner, and the designer.

Task Relationships

Task relationships determine the activities that must occur before, after, or simultaneous to the defined task. These relationships give the schedule its "logic" and enable it to work. Task relationships also determine the interaction of the parties performing the work.

Phases

A *phase* is a grouping of related tasks. For example, the related tasks that comprise a phase for placing a brick wall are as follows:

- purchase sand, masonry cement, and brick
- rent mortar mixer and scaffolding
- place brick
- clean brick

Dividing your schedule into phases enables you to focus on one aspect of the project at a time, determine if any tasks are missing from the phase, and report the schedule status at the right level of detail.

Milestones

Milestones are tasks that do not represent actions. You can use milestones as interim goals to track the progress of the project. An example of a milestone would be having a house "dried in" to obtain a progress payment.

Gantt Charts and Reports

The *Gantt Chart* provides a convenient and easy-to-read method for viewing the schedule (Fig. 1.1). Simply stated, the horizontal axis represents the timescale, and the vertical axis lists the tasks. Each task's duration is shown graphically in relation to the timescale to show when the task is planned to be executed. You can make the task names as broad or narrow as you need to adequately describe the project.

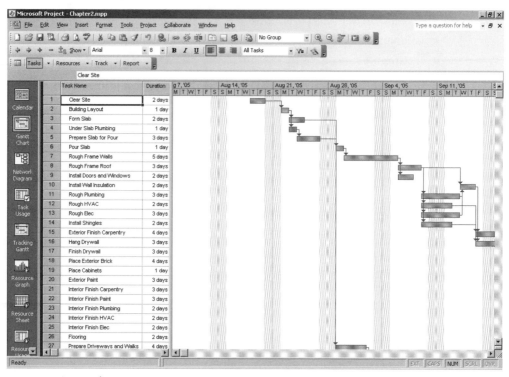

Figure 1.1 Gantt Chart

Rough Logic Diagram

Preparing a rough logic diagram is the first step of the planning phase. You will use the rough logic diagram to build your project on paper. The diagram defines tasks and the relationships between the tasks. Once the rough logic diagram has developed, refined, and accepted by all relevant parties, it becomes your project plan.

Preparing the Rough Logic Diagram

The main steps in preparing a residential schedule are as follows:

1. **Initial meeting.** You should convene an initial meeting with your project team, which includes the superintendent, foreperson, estimator, trade con-

tractors, fabricators, suppliers, vendors, and designers. The group should use the estimate to build the project on paper. Be sure to designate someone to record the meeting minutes.

2. **Gather the pertinent information.** You will need to gather the following types of information:

   - owner time constraints and other input
   - scope definition
   - building methods and procedures to be used
   - productivity rates, crew balances, and crew sizes
   - construction equipment to be used
   - construction equipment availability
   - material availability
   - trade contractor availability
   - fabricator availability
   - temporary facilities requirements
   - permits, test requirements

3. **Develop the rough logic diagram.** Use the information from the initial meeting to draw the rough logic diagram on paper and to show interrelationships between the tasks. Some builders prefer to use Microsoft® Project rather than manually drawing the rough logic diagram on paper.

4. **Develop the schedule.** Add task duration data and, where possible, begin to list resources and costs to the plan.

5. **Review the schedule.** The entire project team must accept the rough logic diagram. It is critical for the team to agree with your concept, methods, and procedures to be used in the construction of the project.

6. **Get buy-in from major trade contractors and suppliers.** As the home builder, you are responsible for coordinating and scheduling the work and resolving any disputes. By meeting with the trade contractors and suppliers in advance, you can modify the plan and reconcile differences. Be sure that you obtain information from any parties whose work can impact the schedule.

7. **Accept project schedule.** Once you review the rough schedule with the trade contractors and suppliers, and they sign-off on it, it becomes the official project schedule.

Criteria for Success

Each home building project has a start date, an end date, and a unique set of characteristics and tasks that must be accomplished in order to complete the project. Project success is usually judged by the following preset criteria:

- Did the project come in on budget?
- Did the project come in on time?
- Were your clients and employees satisfied?
- Was this project an efficient use of company resources?

Next we'll explore creating a schedule using Microsoft® Project.

CHAPTER 2

Creating the Microsoft® Project Schedule

Objectives

The purpose of this chapter is to familiarize you with the basic Microsoft® Project attributes and capabilities and to begin the scheduling process. Use the Chapter2.mpp file on the CD to follow along and perform the practice skills. After completing this chapter, you will be able to:

- create and open project files
- save files
- identify the Gantt Chart
- utilize views
- adjust timescales
- add tasks
- enter task name
- enter task length
- enter milestones
- enter task links
- enter task constraints
- identify the critical path

Creating and Opening Microsoft® Project Files

When you launch Microsoft® Project, a blank project file is automatically displayed in the Gantt Chart view (Fig. 2.1). To open an existing project, select the File menu and click Open to select the file. To start a new project, select the File menu and click New (Fig. 2.2).

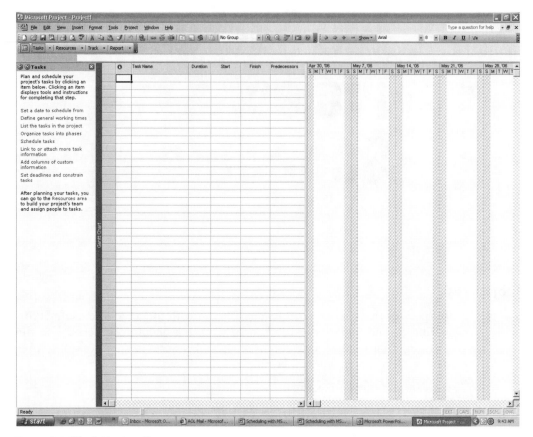

Figure 2.1 Blank Project Screen

The New Project tab will appear on the left side of the screen (Fig. 2.3). The New Project tab provides preconfigured templates for creating new schedules. Click blank project and the Tasks tab for the newly created schedule will appear (Fig. 2.4).

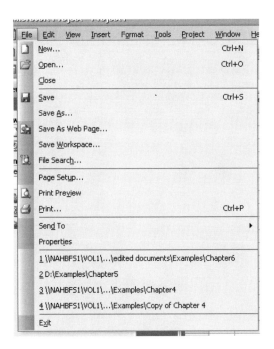

Figure 2.2 File Menu

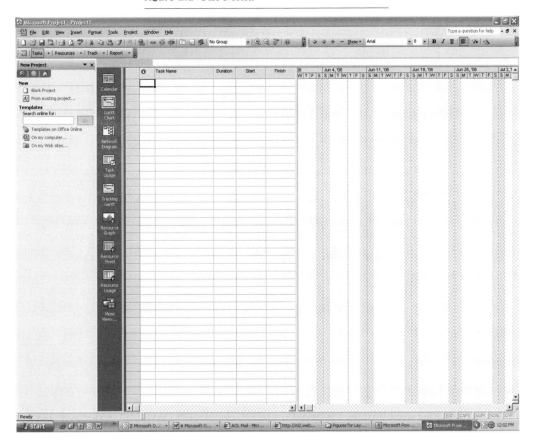

Figure 2.3 Create New Project

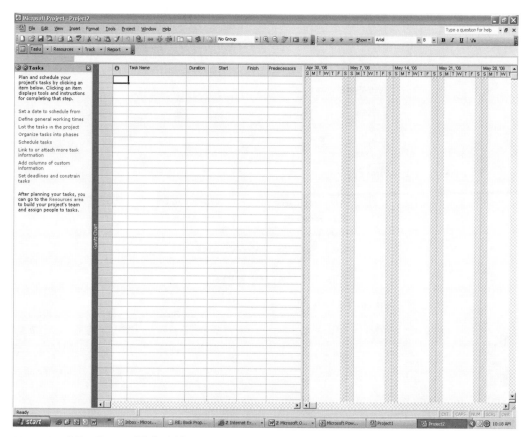

Figure 2.4 New Project Tasks Tab

PRACTICE THIS SKILL

1. To enter the project start date, click Set a date to schedule from. The Set a date to schedule from tab will appear on the left side of the screen (Fig. 2.5).

2. To select the month and year the project will start, click the down arrow (located next to the blank date field).

3. Select 1/1/07 as the date the project will start.

4. An empty project file appears (Fig. 2.4). The Gantt Chart view is displayed by default.

Saving the File

When you are building a schedule, you should always save your changes. Select the File menu and click Save (Fig. 2.2). If this is the first time you have saved the sched-

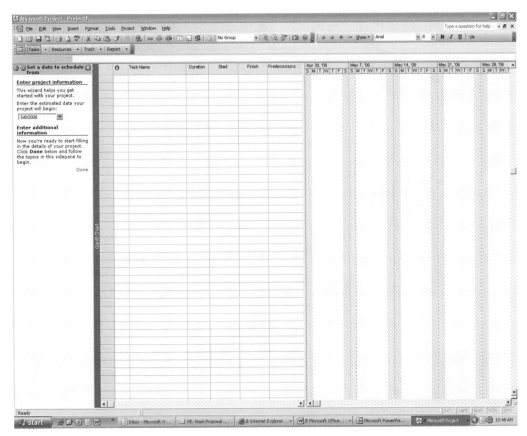

Figure 2.5 Set Project Start Date

ule, the Save As dialog box will appear (Fig. 2.6). To save the project you just created, type "Chapter 2 Exercise" in the File name field and click Save. (We will not save the changes we make to the files on the practice CD, as they are for practice purposes only).

Identifying the Gantt Chart

The Gantt Chart is one of the most familiar tools for visualizing a home building project (Fig. 2.7). Each task or activity is represented as a single horizontal bar. These task bars are positioned across a period of time called a *timescale*, which is displayed at the top of the chart. The length of an individual task bar represents a task's duration, or the time it takes to complete that task. For example, in Figure 2.7, the Building Layout task cannot begin until its predecessor, Clear Site, is complete. This basic arrangement is an excellent tool for quickly assessing home building tasks over time. *Link lines* connecting task bars reflect relationships between tasks, such as not being able to start a given task until another one is finished.

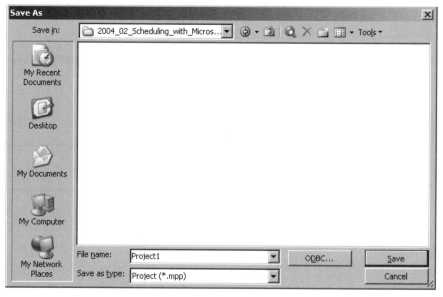

Figure 2.6 Save As Dialog Box

	ⓘ	Task Name	Duration	Aug 14, '05	Aug 21, '05	Aug 28, '05	Sep 4, '05
1		Clear Site	2 days				
2		Building Layout	1 day				
3		Form Slab	2 days				
4		Under Slab Plumbing	1 day				
5		Prepare Slab for Pour	3 days				
6		Pour Slab	1 day				
7		Rough Frame Walls	6 days				

Figure 2.7 Gantt Chart

Views

A *view* is the format in which you can enter and display information in Microsoft® Project. The Gantt Chart is the default view (Fig. 2.4). The Gantt Chart is comprised of the entry table on the left, which is used to enter task information, and the bar chart on the right, which graphically represents the task information on a timescale. The timescale is the time period indicator at the top of a given view.

A single view is simply a single sheet, chart, graph, or form. A combination view displays any two singles together (for example, the Gantt Chart combines the entry table sheet view and the bar chart.) Microsoft® Project has three different types of views:

- **Sheet view.** Displays task or resource information in a row and column format. Sheet views are useful for entering or viewing a lot of information at one time.
- **Chart/graph view.** Provides a graphical representation of your tasks or resources. Chart/graph views are useful when you need to visually present information without all the details.
- **Form view.** Displays individual task or resource information. Form views are useful when you want to focus on detailed information about a specific task or resource.

You can display each view by using the View menu or the View bar (Fig. 2.8). To display the View bar, select the View menu and click View bar. The View bar displays eight of the most commonly used views as icons, plus an icon to display additional views. The eight most commonly used views are as follows:

1. **Calendar.** Shows the project tasks and the duration of those tasks by month. Use this view to show the tasks scheduled in a specific week or range of weeks.
2. **Gantt Chart.** Provides a list of tasks and related information and a chart showing tasks and their duration over time. Use this view to enter and schedule a list of tasks.
3. **Network Diagram.** Shows all tasks and task dependencies. Use this view to create and fine-tune your schedule in a flowchart format.
4. **Task Usage.** Provides a list of tasks with the assigned resources grouped under each task. Use this view to see which resources are assigned to specific tasks and to view resource work contours.
5. **Tracking Gantt.** Provides a list of tasks with related information and a chart showing baseline and scheduled Gantt bars for each task. Use this view to compare the baseline schedule with the actual schedule.
6. **Resource Graph.** Shows resource allocation, cost, or work over time. Use this view to display information about a single resource or group of resources.
7. **Resource Sheet.** Provides a list of resources and related information. Use this view to enter and edit resource information in a spreadsheet format.
8. **Resource Usage.** Provides a list of resources that show task assignments grouped under each resource. Use this view to show cost or work allocation information over time for each resource per assignment and to set resource work contours.

You can access a list of every available view by opening the More Views dialog box.

Figure 2.8   View Bar

Adjusting the Timescale

In addition to moving within a view to display additional project information, you can also adjust the timescale in a chart view to display additional graphical information (Fig. 2.9). To access the timescale dialog box, select Timescale from the Format menu. The timescale is located across the top of the Gantt Chart. It represents the time when a task takes place. The timescale includes a top tier, a middle tier, and a bottom tier. The top tier displays large units of time, the middle tier displays small units of time, and the bottom tier displays the smallest units of time.

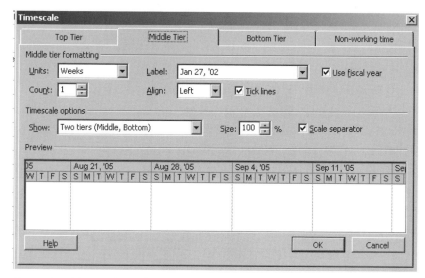

Figure 2.9 Timescale Dialog Box

PRACTICE THIS SKILL

1. Open the Chapter2.mpp file on the CD.
2. Use the Zoom In and Zoom Out buttons located on the toolbar to adjust the timescale tiers. The Zoom In button decreases the timescale into smaller units (down to hours over 15-minute increments) to give you a more detailed view. The Zoom Out button increases the timescale to larger units (up to years over 6-month increments) to give you a broader view.
3. Now try adjusting the timescale using the Timescale dialog box on the Format menu.
4. Close the file without saving your changes.

Entering Task Names

Use the project file to input a list of tasks necessary to complete home construction. You must enter each task in the task Name column of the Entry table in the Gantt Chart view (Fig. 2.10). You can also enter tasks in other views that have a Task Name column. Each task in the task list is associated with a Task Identification (ID) number (Fig. 2.11). As you enter tasks, the Task ID number is automatically assigned and entered in the gray row headings to the left of the Gantt Chart view. When you edit the task list the Task ID numbers are automatically renumbered to keep the list in numerical order. To delete a task, simply select the task to be deleted and press Delete on your keyboard, or go to the Edit menu, and click Delete Task.

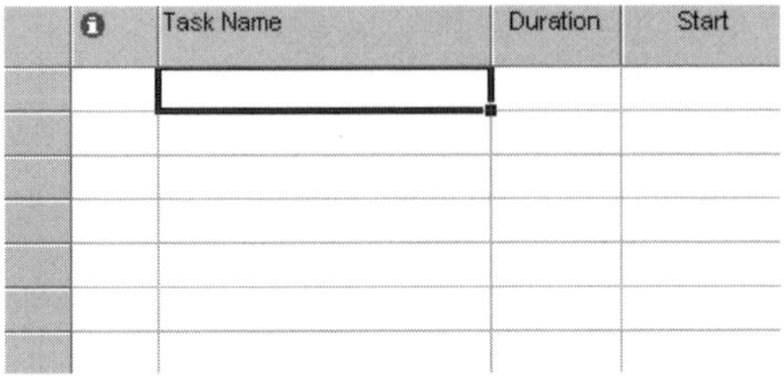

🛈	Task Name	Duration	Start

Figure 2.10 Task Name Column

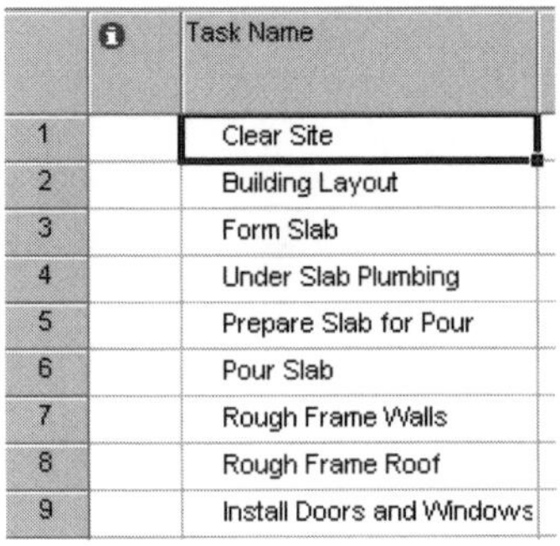

	🛈	Task Name
1		Clear Site
2		Building Layout
3		Form Slab
4		Under Slab Plumbing
5		Prepare Slab for Pour
6		Pour Slab
7		Rough Frame Walls
8		Rough Frame Roof
9		Install Doors and Windows

Figure 2.11 Task Identification (ID) Numbers

PRACTICE THIS SKILL

1. Using the file you created earlier, Chapter 2 Exercise, enter the following tasks:

   a. clear site

   b. building layout

   c. form slab

   d. under slab plumbing

    e. prepare slab for pour

    f. pour slab

2. Delete the "prepare slab for pour" task.

3. Save the file.

Entering Task Duration

You must enter a duration estimate, or the amount of time it will take to accomplish the task, for each task (Fig. 2.12). You can specify duration in values of minutes, hours, days, or weeks as *working* time or *elapsed* time. A unit of working time is confined by the hours of the day and the number of days that resources are actually being used. A unit of elapsed time includes both working and *nonworking* (idle) time based on a 24/7 schedule. One day (1 day) is the default duration time. As you enter tasks, the default duration estimate is automatically entered in Duration column. You can change the default duration by entering a new value and unit of time in the Duration field.

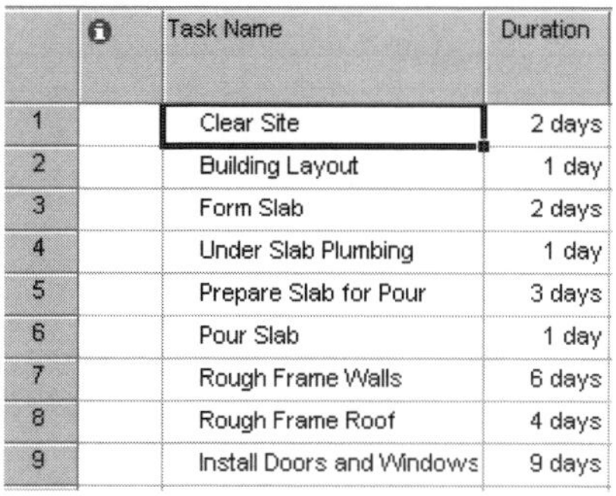

	❶	Task Name	Duration
1		Clear Site	2 days
2		Building Layout	1 day
3		Form Slab	2 days
4		Under Slab Plumbing	1 day
5		Prepare Slab for Pour	3 days
6		Pour Slab	1 day
7		Rough Frame Walls	6 days
8		Rough Frame Roof	4 days
9		Install Doors and Windows	9 days

Figure 2.12 Task Duration Estimates

PRACTICE THIS SKILL

1. Using the file you created earlier, Chapter 2 Exercise, enter the following task durations:

    a. clear site: 2 days

    b. building layout: 1 day

    c. form slab: 2 days

    d. under slab plumbing: 1 day

    e. pour slab: 1 day

2. Save the file.

You can enter task information by selecting a field in the Entry table or by selecting the Task ID heading. By default, the cursor moves one row down when you press the Enter key inside a field. When you press Tab, the cursor moves one field to the right. When you press Shift+Tab, the cursor moves one field to the left. The cursor will continue to cycle through the selected row until you select another field. You can also use the mouse or the keyboard arrows to move from field to field in the Entry table.

As you develop your schedule and perform the home building tasks, you may discover that you need to add new or additional tasks.

PRACTICE THIS SKILL

1. Using the file you created earlier, Chapter 2 Exercise, select the Pour Slab task row.
2. Click the Insert menu and select New Task.
3. Type the new task name "Prepare Slab for Pour."
4. Enter a duration of 3 days for the new task.
5. Save your changes.

You can drag and drop tasks within the task list (through selection of the Task ID heading). Microsoft® Project automatically inserts a row at the new location and deletes the row in the prior location. You can also use the Cut, Copy, and Paste functions to move tasks, but you must use the Task ID heading in order to affect the entire row.

Entering Recurring Tasks

A task that occurs repeatedly is called a *recurring* task (Fig. 2.13). A recurring task might be a weekly meeting, status report, or regular inspection. You can create a recurring task instead of retyping the task and duration. Use the Recurring Task Information dialog box to specify the parameters of the recurring task.

PRACTICE THIS SKILL

1. Open Chapter2.mpp on the CD.
2. Select the Clear Site task row.
3. Click the Insert menu.
4. Select Recurring Task.
5. Type "Staff Meeting."
6. Set the task to occur weekly.
7. Specify the recurrence pattern to occur on Monday.
8. Click OK.
9. Close the file without saving your changes.

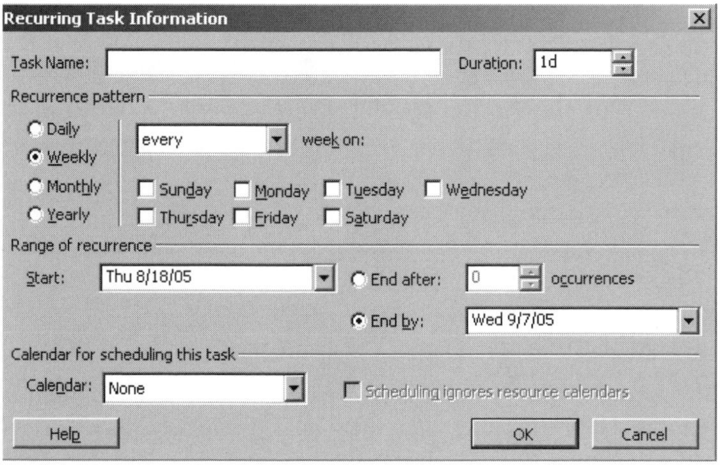
Recurring Task Information
Task Name:
Duration: 1d
Recurrence pattern
Daily
Weekly
Monthly
Yearly
every
week on:
Sunday
Monday
Tuesday
Wednesday
Thursday
Friday
Saturday
Range of recurrence
Start: Thu 8/18/05
End after: 0 occurrences
End by: Wed 9/7/05
Calendar for scheduling this task
Calendar: None
Scheduling ignores resource calendars
Help
OK
Cancel

Figure 2.13 Recurring Task Information Dialog Box

Entering Milestones

As discussed in Chapter 1, milestones represent the completion of an event, phase, or other measurable goal within the home building project. You can add milestones to mark the completion or beginning of significant sections of your home building project. Milestones are tasks that have been assigned a duration of zero (0). When a task becomes a milestone, the bar changes to a diamond-shaped marker with the date the milestone occurs to the left of the marker (Fig. 2.14).

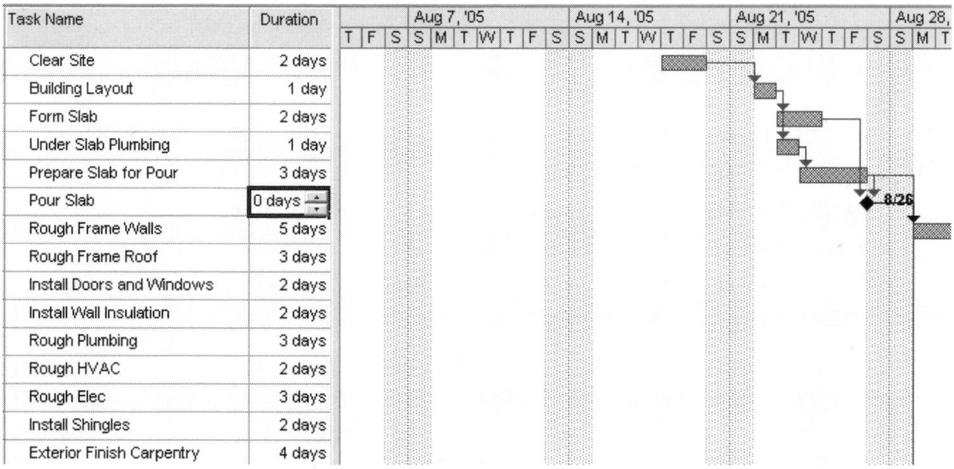
Task Name
Duration
Aug 7, '05
Aug 14, '05
Aug 21, '05
Aug 28,
Clear Site 2 days
Building Layout 1 day
Form Slab 2 days
Under Slab Plumbing 1 day
Prepare Slab for Pour 3 days
Pour Slab 0 days
Rough Frame Walls 5 days
Rough Frame Roof 3 days
Install Doors and Windows 2 days
Install Wall Insulation 2 days
Rough Plumbing 3 days
Rough HVAC 2 days
Rough Elec 3 days
Install Shingles 2 days
Exterior Finish Carpentry 4 days
8/26

Figure 2.14 Milestones

PRACTICE THIS SKILL

1. Using the file you created earlier, Chapter 2 Exercise, select the Pour Slab task.
2. Enter zero (0) in the Duration field.
3. Notice the diamond-shaped milestone marker and date on the Gantt Chart.

Entering Task Links

When you initially enter a task, Microsoft® Project automatically schedules it to begin on the project start date. By linking tasks, you can establish a dependency that determines the sequence of tasks (Fig. 2.15). Microsoft® Project then schedules the tasks by setting the start and finish dates for each task, moves the Gantt task bars in the Gantt Chart view to the appropriate date on the timescale, and draws *link lines* to display the dependency.

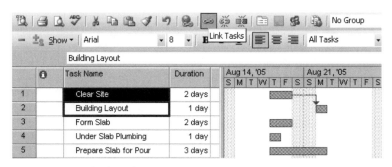

Figure 2.15 Linked Tasks

There are four types of task dependencies: *finish-to-start, finish-to-finish, start-to-start,* and *start-to-finish.*

- **Finish-to-start.** The next task starts as the previous task is completed. This is the most common task dependency.
- **Finish-to-finish.** Both tasks may finish at the same time.
- **Start-to-start.** Both tasks may start at the same time.
- **Start-to-finish.** The start of one task depends on the finish of another task. This is the least common task dependency.

A task that must start or finish before another task can begin is called a *predecessor* task. A task that depends on the start or finish of a preceding task is called a *successor* task. Each dependency can either lengthen or shorten the project schedule duration. For example, a finish-to-start dependency extends the duration because one task must finish before the other can start. The start-to-start and finish-to-finish dependencies can shorten the duration because they overlap the duration of the tasks.

Linking Tasks to Create Task Dependencies

Linking tasks creates a default *finish-to-start dependency* (Fig. 2.16). Once you have entered tasks in the default dependency type, you can begin to identify and address the tasks that are an exception to the common finish-to-start dependency. You can unlink tasks and phases that are not related, and you can link tasks that are not listed consecutively in the task list. You can also link tasks to a single predecessor and successor or to multiple predecessors and successors. You can then unlink the noncontiguous tasks. When a link between tasks is removed, the task, or subtasks within a group, will automatically move back in time.

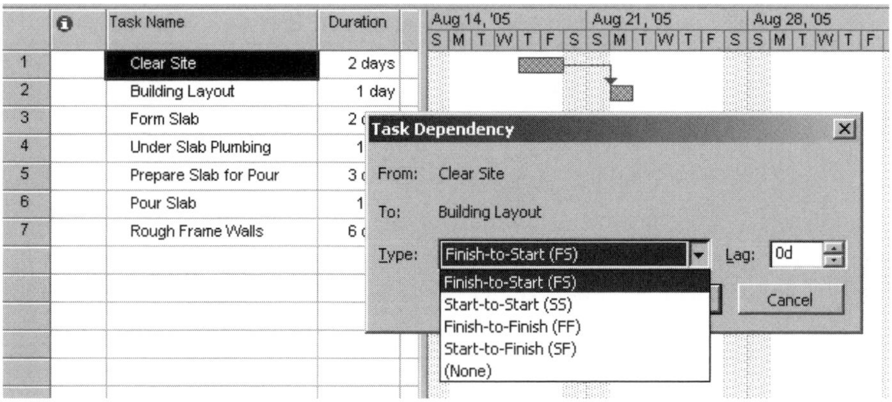

Figure 2.16 Finish-to-Start Dependency

PRACTICE THIS SKILL

1. To link tasks: select the tasks to be linked and click the Link Tasks button on the Standard toolbar (Fig. 2.17).
   - Open the Chapter2.mpp file.
   - Link the following tasks: clear site, building layout, form slab, and under slab pouring.
2. To unlink tasks: select the tasks to be unlinked and click the Unlink Tasks button on the Standard toolbar.
   - Unlink the tasks you just linked.

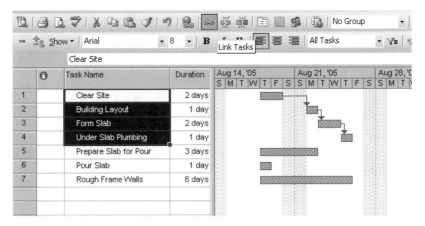

Figure 2.17 Linking Tasks

3. To link noncontiguous tasks: select the first task to be linked and press Ctrl, then select the next task to be linked, and click the Link Tasks button on the Standard toolbar.

   ● Link the following tasks: clear site and under slab pouring.

You can also drag and drops tasks to be linked:

1. Place the mouse pointer over the center of the Gantt bar for the first task in the link (the pointer changes to a four-headed arrow).

2. Drag down to the center of the Gantt bar for the second task in the link. (As you drag over the task bars, the pointer changes to a chain link and a Screen Tip indicating the link will be established is displayed.)

3. Release the mouse button when the Screen Tip displays the correct information.

4. Close the file without saving your changes.

Changing Task Dependencies

To change the dependency type, use the Task Dependency dialog box or any view that displays the Type field. You can display the Task Dependency dialog box by double-clicking a link line between tasks (Fig. 2.16). The dialog box confirms which tasks the link line is connected to and displays the current dependency. The Type box is used to change the dependency. When a task dependency is changed, Microsoft® Project redraws the link line on the Gantt Chart to reflect the change. Link lines allow you to quickly identify task dependencies.

Task Form

You can also view information about task dependencies in the Task Form located under the View menu in the option More Views (Fig. 2.18). The Task Form pro-

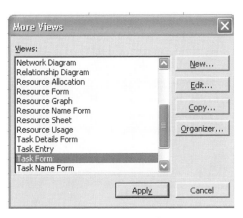

Figure 2.18 More Views Dialog Box

vides details about an individual task, such as its start and finish date, predecessor, and dependency type. The Task Form can be viewed as a single view, or you can select the Task Entry view to see a combination of the Task Form View and the Gantt Chart view.

You can view, enter, and edit basic tracking and scheduling information about individual tasks in the Task Form (Fig. 2.19). Use the Task Form when you want to display detailed information about a task or to enter and revise task information in one location.

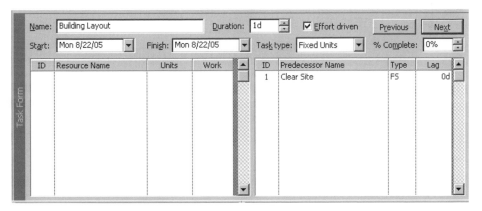

Figure 2.19 Task Form View

Lead Time and Lag Time

In addition to changing the dependency type, you can further define the true impact of task relationships using *lead* time and *lag* time (Fig. 2.20). You can access the Task Dependency dialog box by double-clicking the relationship line between the two selected activities. Lead time creates an overlap in a task dependency that can shorten the project duration. For example, if you specify a lead time of one day on a finish-to-start dependency, the two tasks overlap in time by one day. In other words, the last day of the first task takes place while the first day of the second task takes place. Lag time creates a delay, or gap, in the task dependency that can lengthen the project duration. For example, if you specify a lag time of one day on a finish-to-start dependency, there is a one-day gap between the tasks. In other words, the first task finishes, a day goes by, and the second task starts. Lead time moves the start of the successor task back in time, and lag time moves the start of the successor task forward in time.

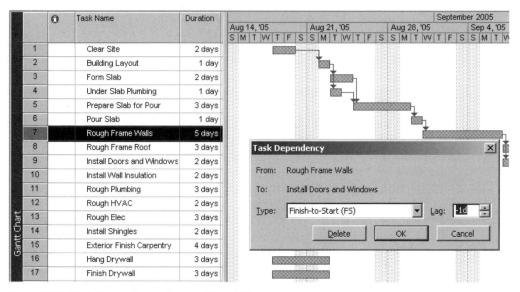

Figure 2.20 Task Dependency: Lag and Lead Time

The Lag field is used to specify both lead and lag. Lead is displayed as a negative number because the project duration is shortened. In contrast, lag is displayed as a positive number because the project duration is lengthened. You can enter lead time and lag time in the Lag box on the Task Dependency dialog box or in any view that displays the Lag field.

PRACTICE THIS SKILL

To specify lead time:

1. Open Chapter2.mpp on the CD.

2. Double-click the link line between Rough Frame Walls and Install Doors and Windows.

3. Press Tab to move to the Lag box.

4. Type "−1" to reflect the amount of lead time.

5. Click OK.

6. Notice how the link line between the tasks has moved back in time.

To specify lag time:

1. Double-click the link line between the previous tasks.
2. Press Tab to move to the Lag box.
3. Type "2" to reflect the amount of lead time.
4. Click OK.
5. Notice how the link line between the tasks has moved forward in time.
6. Close the file without saving your changes.

Entering Task Constraints

Microsoft® Project defaults all task start dates with the start date of the home building project, and likewise, all task finish dates with the project finish date. By linking and assigning resources, the program sets start and finish dates according to their task dependencies. You can take these parameters one step further by applying either *flexible* (not date specific) or *inflexible* (date specific) constraints on specific tasks.

The Microsoft® Project default is the As Soon As Possible constraint. It is important that you realize when to deviate from the default constraint. For example, an often-used resource might need to be assigned an "As Late As Possible" constraint on less critical tasks so that it will be available when needed for critical path tasks. Other constraints include: Start No Earlier Than, Finish No Earlier Than, Start No Later Than, and Finish No Later Than.

PRACTICE THIS SKILL

1. Open Chapter2.mpp on the CD.
2. Press F5.
3. Type "8" in the ID field.
4. Double-click the Rough Frame Roof task.
5. Check the finish date in the General tab in the Task Information Dialog Box.
6. Click the Advanced tab (Fig. 2.21).
7. Click the drop-down arrow in the Constraint type field and select As Soon As Possible.
8. Click the drop-down arrow in the Constraint date and select a date that is two days before the finish date.
9. Click OK.
10. Close the file without saving your changes.

Once a constraint has been applied to a certain task, an icon will appear in the Indicators field. You can use your mouse to point at these icons to display specific task constraint types and dates. To view task constraints, click the More Views icon on the View bar, click Task Sheet in the Views dialog box, and then click Apply.

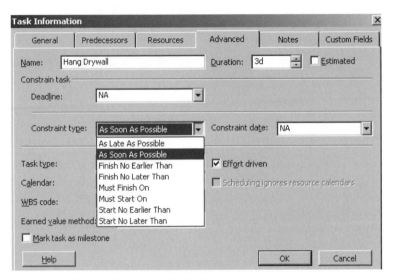

Figure 2.21 Task Information Dialog Box

Identifying the Critical Path

The *critical path* identifies those tasks that are critical to the duration of the project. A *critical task* cannot have its duration lengthened or its start date delayed without impacting the project finish date. Critical tasks do not have slack time. Critical tasks form a critical path through the home building project. Once a residential schedule is formatted to display the critical path, you can reduce or lengthen the total project duration by changing the duration, dependencies, or resources of the critical tasks.

Use the Gantt Chart Wizard button from the toolbar to format the Gantt Chart to automatically display the critical path (Fig. 2.22). When the Gantt Chart is formatted to display the critical path, critical task bars are displayed as red, and noncritical task bars are displayed as blue (Fig. 2.23). The critical path will be automatically updated as you make changes to the project schedule.

PRACTICE THIS SKILL

1. Open Chapter2.mpp on the CD.

2. Click the Gantt Chart Wizard button on the Standard toolbar.

3. Click Next.

4. Select the Critical Path option and click Next.

5. Select the default selection and click Next.

6. Select Yes and click Next.

7. Click Format It and then click Exit Wizard.

8. Close the file without saving your changes.

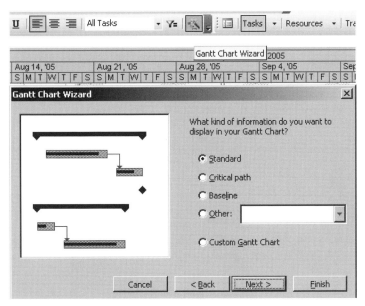

Figure 2.22 Gantt Chart Wizard Dialog Box

		Task Name	Duration
1		Clear Site	2 days
2		Building Layout	1 day
3		Form Slab	2 days
4		Under Slab Plumbing	1 day
5		Prepare Slab for Pour	3 days
6		Pour Slab	1 day
7		Rough Frame Walls	5 days
8		Rough Frame Roof	3 days
9		Install Doors and Windows	2 days
10		Install Wall Insulation	2 days
11		Rough Plumbing	3 days
12		Rough HVAC	2 days
13		Rough Elec	3 days

Figure 2.23 Gantt Chart Showing Critical Path

Entering Tasks: Sort/Filter/Group

Sort

The tasks in Figure 2.23 are sorted by ID number. Microsoft® Project gives you the ability to sort by the following options:

- start date
- finish date
- priority
- cost
- ID#

To change the sort criteria, select the Project menu and click the Sort option (Fig. 2.24).

Select the Sort by option to identify the sort and level options. You can input up to three levels of sort criteria.

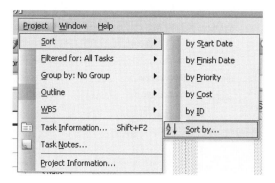

Figure 2.24 Sort Options

Filter

Microsoft® Project allows you to filter tasks so that all tasks do not appear on the screen. To set a filter criterion, select the Project menu and click Filter for: (Fig. 2.25). You will see the following options:

- All Tasks (default)
- Completed Tasks
- Critical
- Date Range
- Incomplete Tasks
- Milestones
- Summary Tasks
- Task Range

- Tasks With Estimated Durations
- Using Resource…

You can access other filter options by selecting More Filters from the Filtered for options.

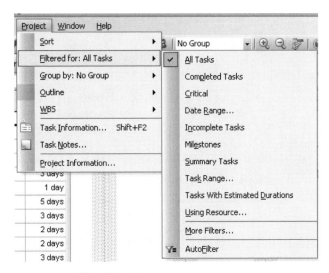

Figure 2.25 Filter For: Options

Group

Microsoft® Project gives you the ability to group tasks by criteria, which can be a real advantage when you want to place tasks with common features together. To select a group criterion, select the Project menu and click Group by: (Fig. 2.26). You will see the following options:

- No Group (default)
- Complete and Incomplete Tasks
- Constraint Type
- Critical
- Duration
- Duration Then Priority
- Milestones
- Priority
- Priority Keeping Outline Structure

You can access other filter options by selecting the More Groups option from the Group by: options.

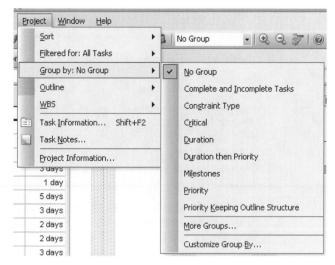

Figure 2.26 Group By: Options

PRACTICE THIS SKILL

1. Open Chapter2.mpp.
2. Sort the schedule by Start Date.
3. Filter the schedule by Incomplete Tasks.
4. Group the schedule by Constraint Type.
5. Close the file without saving your changes.

Schedule Options

The Options dialog box is one of the most important features in Microsoft® Project because it allows you to control your schedule's environment (Fig. 2.27). By using the Options dialog box, you can change several aspects of your schedule, including the default view, how calculations are performed, and spelling rules, just to name a few. You can access the Options Dialog Box by selecting Options from the Tools menu. You will see the following tabs:

- View
- General
- Edit
- Calendar
- Schedule
- Calculation

- Spelling
- Save
- Interface
- Security

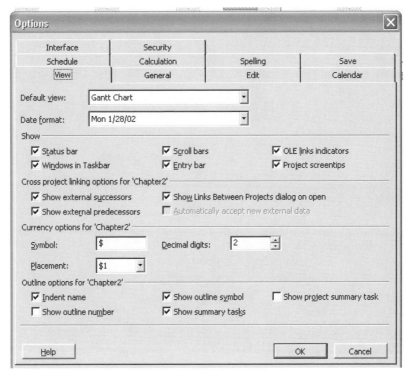

Figure 2.27 Options Dialog Box

Now that you understand the basics, let's add resources and costs to the schedule.

CHAPTER 3

Resources and Costs

Objectives

This chapter will teach you how to use Microsoft® Project to manage your resources and costs. Use the Chapter3.mpp file on the CD to follow along and perform the practice skills. After completing this chapter, you will be able to do the following:

- create a resource list
- assign costs to the resource list
- assign resources/costs to tasks
- add resource notes

Controlling Resources and Costs

You have a great deal of flexibility in determining the resource requirements of your residential construction project. You must establish the labor, material, and, possibly, the construction equipment requirements of each task in order to evaluate and refine the requirements of the project. Remember, time is money. You should determine what resources you will need based on the efficient use of time, money, and the resources themselves.

Resources

You must control all of your project's resources. Controlling resources involves managing labor hours, bulk materials, construction equipment, and permanent equipment in a manner that is time- and cost-effective. Your ability to get the greatest "bang for the buck" in controlling resources will determine your project's success. To successfully control your resources, you must be detail-oriented and you must maximize your resource utilization, monitor costs and time, and control waste.

Resource loading. You can effectively control your resources by loading the workers, materials, and construction equipment you plan to use in Microsoft® Project. By inputting and assigning the labor resources in advance you can identify which tasks must be accomplished on a particular day, as well as the number of workers per craft that are required. Breaking down the home building craft-type work allows you to effectively measure construction progress and performance.

Resource limits. A resource limit is the maximum amount of a resource that is available at once. Therefore, if a person serves a unique function on a project, the limit is one unit. On the other hand, if five workers of one craft are available at any one time, then the limit is five units. Obviously, resource limitations impact the scheduling of the project and the task relationships.

Resource leveling. Resource leveling redistributes resources to eliminate resource conflict. If a resource is over allocated, Microsoft® Project can reschedule the activities so that the resources are not overcommitted. In other words, the project is rescheduled so that the resource does not exceed its limit, and it is therefore leveled.

Costs

The most important resource is money. Scheduling and controlling the expenditure of funds is critical to the building process. Determining how much money you will need for each task at the right time can be tricky business. Ultimately, the owner pays for a project, but you must make sure that money is available to finance interim periods until the monthly payment from the owner is available. With a good estimate, accurate cash projections, and intelligent distribution of funds, you can finance the project without having to borrow funds.

Cash needs. As the builder, you need funds to build your projects and also to run your business operations. You need funds for payroll, materials, supplies, and construction equipment.

Timing of expenditures. To finance projects, you need to maintain a positive cash flow without having to borrow funds or take cash from interest earning accounts. The goal is to have each project stand on its own. In addition to distributing and tracking costs by task, you will need to control the following:

- payment request to owner
- labor productivity and costs
- material and supplies cost
- subcontractors
- overhead costs
- payment of funds

Controlling costs. Controlling the expenditure of funds is as important as controlling time. Most home building project contracts call for the builder to be paid monthly or upon completion of preset phases of work. As the builder, you must ensure that the project progresses toward a satisfactory completion for the owner. You must also ensure that funds are being properly spent and that you are being paid according to your progress on the project.

Measuring physical progress. The primary reason you need a schedule is to evaluate the physical progress of the project. Some owners now require a task cost-loaded schedule. A task cost-loaded schedule, as opposed to a schedule-of-values, breaks down costs by specific task. An added benefit of this type of schedule is that it allows you and the owner to reach an agreement on costs at the beginning of the process. Also, the percent complete will less likely be an issue at each pay period, because the completion of a particular task is much easier to judge than progress in broad phases of cost. Both you and the owner stand to benefit when resources and costs are assigned to project tasks in advance.

Creating Resource Lists

You must assign applicable resources to each task to control and ensure that they are successfully completed. A given task may require a single resource, or it may require many resources. Remember that resources include people, equipment, material, and other necessities such as permits.

Like tasks, the names of your resources can be general or detailed. For example, a resource name can be a job title, such as "laborer," or it can include a specific proper name, such as "Mark Bounds, assistant superintendent." It all depends on how much detail you want to include. However, be consistent in your naming style.

Defining Resources

The first step in tracking labor, costs, and home completion is to create a resource list. You can easily add resource information for your resources through the Microsoft® Project Resource Sheet view (Fig. 3.1). Click Resource Sheet from the View bar. Open a blank project file and follow these five steps to enter a resource:

1. **Resource Name.** Type the resource names listed in Figure 3.1 in the Resource Name field.

2. **Type.** All resources must be classified as either *Work* (resources that do something, such as labor or construction equipment) or *Material* (resources that are consumed, such as brick). Note that in Figure 3.1, Laborer is classified as a Work type resource and Concrete is a Material type resource. Classify the resources you just typed to match the type shown in Figure 3.1.

   a. The calculations for the resource definition in the Resource Sheet are different depending upon whether the resource is classified as work or material. If the resource is a material resource, select Material from the Type field and type the quantity unit in the Material Label field. Note that in Figure 3.2, CY (cubic yards) is entered in the Material Label field for the Concrete resource. If you selected Work in the type field, you will not be able to enter a quantity in the Material Label field. Type the quantity in the Material Label field as shown in Figure 3.2.

	🛈	Resource Name	Type
1		Carpenter - CL1	Work
2		Carpenter - CL2	Work
3		Laborer	Work
4		Mason	Work
5		Mixer	Work
6		Concrete Finisher	Work
7		Superintendent	Work
8		Dozer	Work
9		Trailer	Work
10		Masonry Saw	Work
11		Radial Arm Saw	Work
12		Generator	Work
13		Pickup	Work
14		Mortar Mixer	Work
15		Concrete	Material ▼

Figure 3.1 Resource Sheet View

	🛈	Resource Name	Type	Material Label
1		Carpenter - CL1	Work	
2		Carpenter - CL2	Work	
3		Laborer	Work	
4		Mason	Work	
5		Mixer	Work	
6		Concrete Finisher	Work	
7		Superintendent	Work	
8		Dozer	Work	
9		Trailer	Work	
10		Masonry Saw	Work	
11		Radial Arm Saw	Work	
12		Generator	Work	
13		Pickup	Work	
14		Mortar Mixer	Work	
15		Concrete	Material	CY

Figure 3.2 Material Label Field

3. **Material Label.** Type the unit of measurement (SF, CY, Tons) for the particular resource. This label should agree with the measurement used to calculate progress.

4. **Initials.** Type the abbreviation for the particular resource for filtering and editing purposes (C for Carpenter, L for Laborer, D for Dozer, etc.).

5. **Group.** To designate a resource group, type a name for the resource group in the Group field. To add several resources to the same group, type the same group name in the Group field for each resource. Note that in Figure 3.3, a Masonry Crew has been organized in the Group field. The group is comprised of a mason, a mixer, a masonry saw, and a mortar mixer. Type the group name as shown in Figure 3.3.

	ⓘ	Resource Name	Type	Material Label	Initials	Group
1		Carpenter - CL1	Work		C	
2		Carpenter - CL2	Work		C	
3		Laborer	Work		L	
4		Mason	Work		M	Masonry Crew
5		Mixer	Work		M	Masonry Crew
6		Concrete Finisher	Work		C	
7		Superintendent	Work		S	
8		Dozer	Work		D	
9		Trailer	Work		T	
10		Masonry Saw	Work		M	Masonry Crew
11		Radial Arm Saw	Work		R	
12		Generator	Work		G	
13		Pickup	Work		P	
14		Mortar Mixer	Work		M	Masonry Crew
15		Concrete	Material	CY	C	

Figure 3.3 Group Field

6. **Max. Units.** These units indicate the percentage of time a Work type resource will spend on a task. If you assign a person to work full-time on a task, you would type 100% in the Max. Units field. If you assign a person to spend half of his or her time working on a task, then you would type 50% in the Max. Units field. Note that by entering 400% Max. Units for Laborer in Figure 3.4, we have inputted a maximum of up to four laborers assigned to any task. You cannot enter Max. Units for Material Type resources. Enter the Max. Units as shown in Figure 3.4.

7. **Std. Rate/Ovt. Rate.** For a Work Type resource, type the standard rate (cost per period of time) in the Std. Rate field. For example in Figure 3.5, the standard rate for the Laborer is $8/hour and the overtime rate is $12/hour. For a Material type resource, the figure that you input in the Std. Rate field will be the unit cost for the quantity unit listed in the Material Label field. In Figure 3.5, the Concrete resource has quantity units of CY (cubic yards) and a Std. Rate of $50. This means that the resource cost is $50/cubic yard. Enter the Std. Rate and Ovt. Rate as shown in Figure 3.5.

	❶	Resource Name	Type	Material Label	Initials	Group	Max. Units	St
1		Carpenter - CL1	Work		C		400%	
2		Carpenter - CL2	Work		C		400%	
3		Laborer	Work		L		400%	
4		Mason	Work		M	Masonry Crew	400%	
5		Mixer	Work		M	Masonry Crew	400%	
6		Concrete Finisher	Work		C		400%	
7		Superintendent	Work		S		100%	
8		Dozer	Work		D		100%	
9		Trailer	Work		T		100%	
10		Masonry Saw	Work		M	Masonry Crew	100%	
11		Radial Arm Saw	Work		R		100%	
12		Generator	Work		G		100%	
13		Pickup	Work		P		100%	
14		Mortar Mixer	Work		M	Masonry Crew	100%	
15		Concrete	Material	CY	C			

Figure 3.4 Max. Units Field

	❶	Resource Name	Type	Material Label	Initials	Group	Max. Units	Std. Rate	Ovt. Rate
1		Carpenter - CL1	Work		C		400%	$12.00/hr	$18.00/hr
2		Carpenter - CL2	Work		C		400%	$10.00/hr	$15.00/hr
3		Laborer	Work		L		400%	$8.00/hr	$12.00/hr
4		Mason	Work		M	Masonry Crew	400%	$14.00/hr	$21.00/hr
5		Mixer	Work		M	Masonry Crew	400%	$8.00/hr	$12.00/hr
6		Concrete Finisher	Work		C		400%	$12.00/hr	$18.00/hr
7		Superintendent	Work		S		100%	$16.00/hr	$24.00/hr
8		Dozer	Work		D		100%	$10.00/hr	$0.00/hr
9		Trailer	Work		T		100%	$8.00/hr	$0.00/hr
10		Masonry Saw	Work		M	Masonry Crew	100%	$1.00/hr	$0.00/hr
11		Radial Arm Saw	Work		R		100%	$1.00/hr	$0.00/hr
12		Generator	Work		G		100%	$2.00/hr	$0.00/hr
13		Pickup	Work		P		100%	$2.00/hr	$0.00/hr
14		Mortar Mixer	Work		M	Masonry Crew	100%	$2.00/hr	$0.00/hr
15		Concrete	Material	CY	C			$50.00	

Figure 3.5 Std. Rate and Ovt. Rate Fields

You can also use the Calendar field, within the Resource Sheet view, to define the availability of resources by calendar restraints. The Code field allows you to relate a resource to your cost code or accounting system.

The Resource Information dialog box allows you to see all of the pertinent information relating to a resource (Fig. 3.6). You can access this dialog box by right-clicking in the Resource Name field of the Resource sheet.

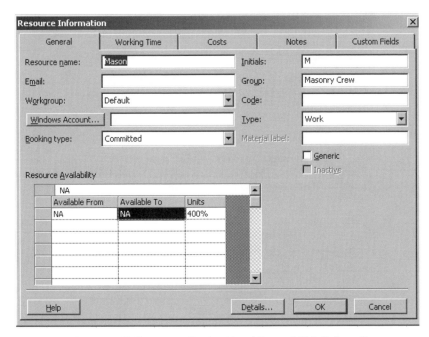

Figure 3.6 Resource Information Dialog Box (General Tab Selected)

Assigning Costs to Resource Lists

Accurate cost information allows you to analyze potential costs for over- and underexpenditures, and ultimately to create and control the home building budget. Before Microsoft® Project can calculate accurate resource costs related to a task, you must specify which cost components and calculation methods to use. There are five cost rate tables (marked A – E) under the Costs tab in the Resource Information dialog box (Fig. 3.7):

- **Rate-based work.** Relates to Work type resources to which you can assign standard cost rates. Microsoft® Project calculates the total resource cost based on the hourly resource rate multiplied by the time it takes to accomplish the task.
  - Rate-based work resource cost = pay rate × time worked
  - Example: cost of mason = $14/hour × 8 hours = $112
- **Rate-based work overtime.** Relates to overtime expenses for Work type resources. For example, the mason in Figure 3.5 has an overtime rate of $21/hour. Microsoft® Project does not automatically calculate costs associated with overtime pay unless you specifically assign the additional hours as overtime work.
- **Rate-based material.** Relates to Material type resources to which you can assign unit cost rates. Microsoft Project® calculates material cost totals based

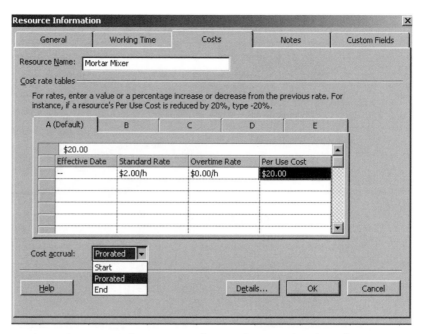

Figure 3.7 Resource Information Dialog Box (Costs Tab Selected)

on the material resource rate multiplied by the quantity of material required to complete the task.

- Rate-based material resource cost = cost/unit × quantity of task units
- Example: cost of concrete = $50/CY × 12 CY = $600

■ **Per use.** A set, one-time fee for the use of either a Work or a Material type resource. The fee is assigned each time the resource is used. You can enter per use costs in addition to a rate-based cost. For example, the mortar mixer in Figure 3.7 has a per use cost of $20. A $20 charge will be added to the $2/hour standard rate every time you use the mortar mixer. This per use cost for the mortar mixer could be for equipment delivery and set up-cost.

■ **Fixed.** The set cost for a task for which you know exactly how much the material will cost. This cost remains constant regardless of the task duration or the work performed by a resource. A rate-based resource cost will increase when the task takes more time, whereas a fixed cost does not.

PRACTICE THIS SKILL

To access the Fixed Cost field:

1. Open Chapter3.mpp on the CD.
2. Select the Gantt Chart from the View bar.

3. Select the View menu.

4. Select Table: Entry (Table: Entry is the default view).

5. Click Cost to view the Fixed Cost field. (Note that in Figure 3.8, the fixed cost for a trade contractor to paint the exterior of the house is $1,500. You can also use this cost type if you don't want to specify a detailed material list per task. In this case, simply enter the total cost of all materials as fixed costs).

6. Close the file without saving your changes.

	Task Name	Fixed Cost
1	Clear Site	$0.00
2	Building Layout	$0.00
3	Form Slab	$0.00
4	Under Slab Plumbing	$0.00
5	Prepare Slab for Pour	$0.00
6	Pour Slab	$0.00
7	Rough Frame Walls	$0.00
8	Rough Frame Roof	$0.00
9	Install Doors and Windows	$0.00
10	Install Wall Insulation	$0.00
11	Rough Plumbing	$0.00
12	Rough HVAC	$0.00
13	Rough Elec	$0.00
14	Install Shingles	$0.00
15	Exterior Finish Carpentry	$0.00
16	Hang Drywall	$0.00
17	Finish Drywall	$0.00
18	Place Exterior Brick	$0.00
19	Place Cabinets	$0.00
20	Exterior Paint	$1,500.00
21	Interior Finish Carpentry	$0.00
22	Interior Finish Paint	$0.00

Figure 3.8 Fixed Cost Field

You can use the Effective Date field to enter different periods for future rate changes, such as pay rate changes or material prices (Fig. 3.7). To enter an effective date, click the Effective Date field and select the date from the pop-up calendar. Then input the new Standard Rate for the new effective date.

Use the Cost Accrual field to determine when you should account for task and resource costs (Fig. 3.7). For accurate cash flow projection, you may want to decide whether costs will accrue on individual tasks at the start, the end, or be prorated over the duration of the resource usage.

Assigning Resources/Costs to Tasks

The Assign Resources dialog box provides the easiest way to create a new resource (Fig. 3.9). You can use this dialog box to enter and assign resources to tasks instead of just entering new resources through the Resource Sheet (Fig. 3.1).To access the Assign Resources dialog, click the Assign Resources button on the standard tool bar.

When using the Assign Resource dialog box to assign resources to a task, simply select a task and then click the applicable resource. The default entered in the Max. Units field is 100%. This is the equivalent of one unit. The percentages can be changed to decimal equivalents, which allow you to manipulate home building resources simply by increasing or decreasing the maximum units. Partial units can be assigned to one task and the rest to a separate task. For example, in Figure 3.9 if you changed the maximum units for "Carpenter CL1" to 50% or .5, then only half of the carpenter's time for the day (4 hours) would be used on this task. The other 4 hours could be used elsewhere. If the task required 2 carpenters, you would input 200% for 2 full days.

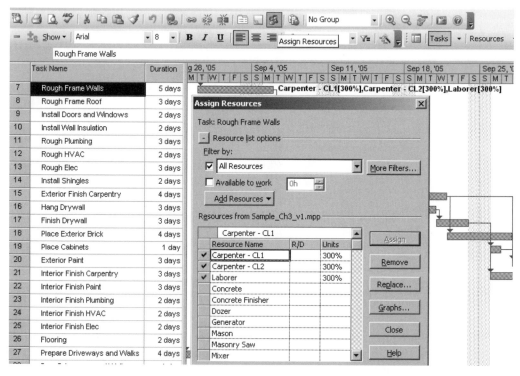

Figure 3.9 Assign Resources Dialog Box

Note that in Figure 3.9, 300% is assigned to Carpenter CL1. This percentage indicates that 3 full-time carpenters are assigned to this task. Also note that "Carpenter – CL1 [300%]" appears beside the task bar on the Gantt Chart. You should ensure that you don't over assign your resources (more units are assigned per resource than there are actually available).

PRACTICE THIS SKILL

1. Open Chapter3.mpp on the CD.

2. Click the Gantt Chart icon on the View bar.

3. Select Rough Frame Walls and click the Assign Resources button on the Standard toolbar.

4. Assign the Mason resource 150% maximum units.

5. Click the Assign button. The task is now assigned. Microsoft Project® automatically draws a checkmark beside the resource name, displays the unit value, and places the resource name next to its respective bar on the bar chart.

6. Close the file without saving your changes.

The Task Information dialog box is another handy feature that allows you to access all task related information (Fig. 3.10). To access this dialog box, select Task Information from the Project menu or double-click on a task. Note that in Figure 3.10, the task information for Rough Frame Walls shows the same three resources that appeared in the Assign Resources dialog box (Fig. 3.9).

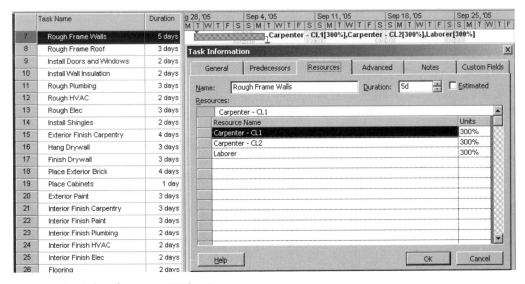

Figure 3.10 Task Information Dialog Box

Recurring Resources

Adding the same resource or resource group to a number of tasks is a real time-saving feature. To use this feature, simply select the tasks that will use the same resource by holding the Ctrl key and clicking the applicable tasks (Fig. 3.11). Then use the Assign Resources dialog box to identify the resource requirements. Note that two carpenters and three laborers were selected in Figure 3.11. As the resource is assigned in the Assign Resources dialog box, the identified resource will appear by the bar for the applicable task on the Gantt chart.

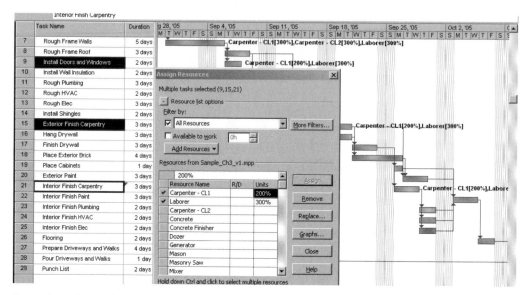

Figure 3.11 Assign Resources Dialog Box

Adding Resource Notes

As with tasks, additional information is often required in order to clarify or document project resources. You can add resource notes by selecting the applicable resource row and right-clicking the mouse (Fig. 3.12). Select Resource Notes to open the Resource Information Notes tab dialog box. Microsoft® Project automatically places a note icon in the indicator field of the Resource Sheet when a note is added.

PRACTICE THIS SKILL

1. Open Chapter3.mpp on the CD.

2. Add the following note to the painter resource: "Be sure to give the revised painting specifications checklist to the painter."

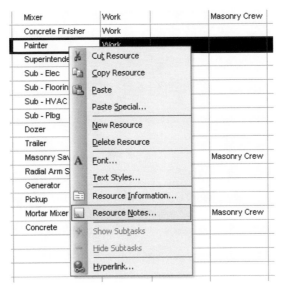

Figure 3.12 Resource Notes

Now that you know how to add resources and costs to your schedule, let's move on to monitoring and updating task durations.

CHAPTER 4

Monitoring and Updating Task Durations

Objectives

You must constantly monitor and update your schedule as the project progresses. This chapter will demonstrate how to update or show actual physical progress for a project. Use the Chapter4.mpp file on the CD to follow along and perform the practice skills. After completing this chapter, you will be able to do the following:

- copy the schedule
- establish baselines
- access project information
- record progress
- change status date
- document changes to the schedule

The Importance of Monitoring and Updating Your Schedule

Why update a home building project? The answer is simple: Very few projects ever go exactly as planned. Among the influences likely to change the original plan are weather, acts of God, productivity (better or worse than anticipated), delivery and labor problems, changes in the scope of the project, interferences between craft or trade contractors, and work flow mismanagement. Therefore, it is necessary to monitor the actual physical progress of the schedule during a building project to keep the schedule current. When you regularly monitor your schedule, you can chart how your work is progressing over time. As the project continues, Microsoft® Project will calculate how your progress will impact the remainder of the tasks to be accomplished.

For a schedule to remain accurate, the actual progress must be continually incorporated into the current schedule. If the update shows that a task was delayed, extended, interrupted, or accelerated, Microsoft® Project will calculate the effects of that change on the successors of the task, and all their successors, all the way through to project completion. When you incorporate actual progress changes into the schedule to make it current, you need to compare the current schedule to the original (or baseline) schedule to determine if you are ahead or behind of your original schedule or goals. To four components to ensuring that your schedule is current are as follows:

- **Develop a baseline schedule.** Before starting the update, you need to save the project *baseline*. The target or original dates represent the baseline schedule.

- **Determine a data date.** You will use this date, which is usually the first day of the new month date, to measure progress. Therefore, at the beginning of each month the current schedule should show the history of what has been completed and what is left to complete. When you compare the current schedule to the baseline schedule, the update should result in a schedule that accurately reflects the current status of the project.

- **Monitor your progress.** The focal point of monitoring a project's progress is determining the status of its tasks. Task status can be labeled as *complete*, *partially complete*, or *no work accomplished*. The tasks that have no physical progress keep their original task relationships and durations. Where there is partial physical progress, the original task relationships are kept and the remaining duration is calculated either in number of days or percent complete. After progress for all applicable tasks is recorded to the data date, the schedule is recalculated. Individual tasks and the entire schedule then can be determined as on, ahead, or behind schedule.

- **Update the schedule.** Because projects are built according to the original plan, documenting changes is key to keeping the building schedule current. The more details you provide about the changes and the more time you devote to analyzing your updates, the better your plans will be. As the construction plan changes, you will need to modify the schedule to show the changes. By changing the schedule to show the current plan, you are creating a living document that shows the current plan throughout the life of the home construction project. To create a current and effective schedule, you should feed your best thinking and information into the schedule on a timely basis.

Copying the Schedule

When updating the schedule, the first thing you should do is copy the previous schedule. Although this is not a requirement, it is good practice. One of the advantages of Microsoft® Project is that it allows you to play "what if" games. It is valuable to have an original version of the schedule that can be recopied and used if the "what if" options do not turn out to be a good idea.

PRACTICE THIS SKILL

1. Open Chapter4.mpp on the CD.
2. Click the File menu.
3. Select the Save As function (Fig. 4.1).
4. Rename the schedule "Chapter 4 Copy."

You can also use the mouse to copy transferable information without the header information. Simply right-click and select Copy Cell. Then select File, New, and Blank Project. Then, right-click in the appropriate cell and select Paste.

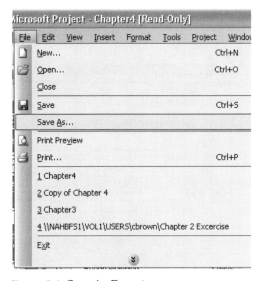

Figure 4.1 Save As Function

Setting Baselines

In creating a current or updated schedule, Microsoft® Project uses the baseline schedule as a comparison when recording actual progress and the expenditure of resources and cash. Setting baselines is critical because it eliminates the need for you to resave a project file when task and resource information changes. You can compare a new baseline to a previous baseline (usually the original schedule) in order to understand how changes affect home construction. Utilizing this updating tool can help you identify potential problems in advance, which enables you to work proactively to solve them. It also enables you to make more precise estimates of task durations and resource requirements as the project proceeds. If you want to compare the baseline to the original schedule before any updated information is input, you should create the baseline before you enter updates.

PRACTICE THIS SKILL

1. Open Chapter4.mpp on the CD.
2. Click the Tools menu.
3. Select Tracking.
4. Click Save Baseline (Fig. 4.2).
5. Click the drop-down menu and select Baseline 1 (you can save up to 11 different baselines by clicking on the Baseline pull-down menu).
6. Select the Save baseline and the Entire project options.
7. Click OK.
8. Close the file without saving your changes.

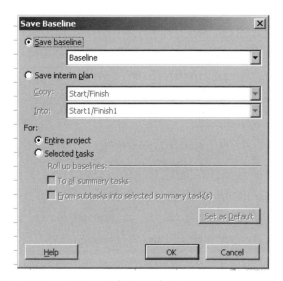

Figure 4.2 Save Baseline Dialog Box

Setting an Interim Plan

After you save the baseline plan and begin updating your schedule you may want to periodically save an *interim plan*. An interim plan is a set of schedule information that is saved at some time during the project. You can compare the interim plan to the baseline in order to monitor project progress.

PRACTICE THIS SKILL

1. Click the Tools menu.
2. Select Tracking.
3. Click Save Baseline (Fig. 4.2).
4. Click the Save interim plan radio button.
5. Select from the Copy pull down menu the fields that need to be copied as the bases for the interim plan. For example: if you want to create an interim schedule based on current start/finish dates, then leave the default options (Start/Finish). However, if you want to create an interim schedule based on a previous interim schedule such as Interim 1, then select Start1/Finish1 in the copy field.
6. From the Into pull down menu, select the fields that will receive the information. For example: if you want to create an interim 2 schedule, then select Start2/Finish2 in the Into field.
7. Close the file without saving your changes.

Adding a Task to a Baseline or Interim Plan

You can add tasks during the scheduling process. If you add a task after you set a baseline or interim plan, you should also add it to the baseline or interim plan. To add a task to the baseline or interim plan, do the following:

1. Click in the Task Name field on the Gantt Chart.
2. Follow steps 2-4 in the Baseline practice instructions above.
3. Click the Save baseline or Save interim plan check box in the Save Baseline dialog box.
4. Click the Selected tasks check box. If you had selected Entire project, rather than Selected tasks, you would have reset the plan for the entire schedule rather that just the new added task.

You would follow the same instructions to change the task baseline information. Simply select the Task on the Gantt Chart and click the Selected tasks check box in the Save Baseline dialog box. Note that when you click Selected tasks, Microsoft® Project only updates the Baseline data for the tasks you selected.

Accessing Project Information
Project Information Dialog Box

The Project Information dialog box allows you to review and edit the information that you entered at the beginning on the project (Fig. 4.3). To view and utilize baseline information, click the Project menu and select Project Information. The

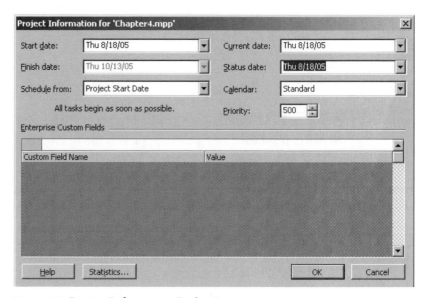

Figure 4.3 Project Information Dialog Box

Project Information dialog box will appear. The Project Information dialog box contains the following key fields:

- **Start date.** Enter a date in this field to schedule your project from the start date. The Finish date field will appear grayed out. Microsoft® Project will calculate the Finish date based upon the task information that you enter.

- **Finish date.** Enter a date in this field to schedule your project from the finish date. The Start date field will appear grayed out. Microsoft® Project will calculate the Start date based upon the task information that you enter.

- **Schedule from.** You have the option to develop your schedule from the Project Start Date or the Project Finish Date.

- **Current date.** This field will contain the current date.

- **Status date.** As of this date, all tasks will be either complete, partial progress will be obtained, or the task will not have been started.

- **Calendar.** The calendar options are 24-Hour, Night Shift, or Standard.

- **Priority.** This number indicates the priority that each activity resource will receive when resources are leveled across multiple projects.

- **Statistics button.** Provides a handy dialog box for assessing the impact of updates and changes on the project (Fig. 4.4).

Recording Progress

You can record your actual progress in one of three ways:

1.   Enter the actual start and finish dates of the task.

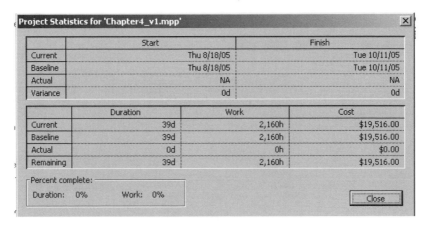

Figure 4.4 Project Statistics Dialog Box

2.   Enter the actual duration of a task.

3.   Indicate progress as a percentage.

Entering Actual Start and Finish Dates for a Task

Entering the actual start and finish dates for a task creates a history of the work as it occurs. You can use this information to recalculate the home construction schedule for the remaining tasks. This historical information can also be valuable in creating schedules for future home building projects.

PRACTICE THIS SKILL

1. Open Chapter4.mpp on the CD.
2. Select the Clear Site task from the Gantt Chart view.
3. Click the Tools menu.
4. Select Tracking.
5. Click Update Tasks. The Update Tasks dialog box for the Clear Site task will appear (Fig. 4.5).
6. Click the down arrow under the Actual Start field to view the pop-up calendar.
7. Notice that, the Clear Site task was actually started on Friday, August 19 (Fig. 4.6). The actual finish date for the task was Monday, August 22. The baseline plan was to start the task on Thursday, August 18 and finish on Friday, August 19 (Fig. 4.5).
8. Click OK.

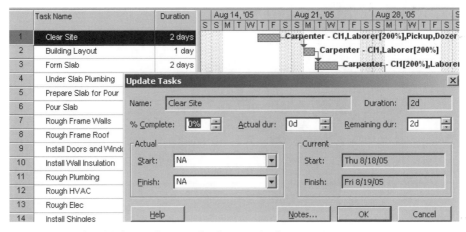

Figure 4.5 Update Tasks Dialog Box (Before Update)

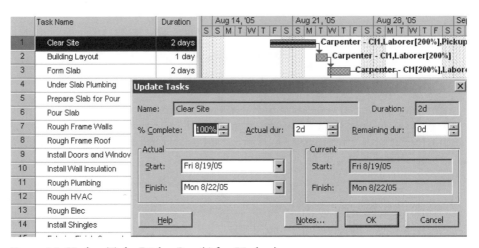

Figure 4.6 Update Tasks Dialog Box (After Update)

Notice the impact of the updated Clear Site task in Figures 4.5 and 4.6. Microsoft® Project automatically changed the following fields:

- **% Complete.** Changed to 100% because the actual finish date was entered, and the task is considered complete.

- **Actual dur.** Changed from 0d to 2d. The actual duration is from Friday, August 19 to Monday, August 22.

- **Remaining dur.** Changed from 2d to 0d because the actual finish date was entered, and the task is considered complete.

The entire schedule has been updated based on the progress of the Clear Site task. The Clear Site task bar on the Gantt Chart now has a dark solid line through it that denotes actual progress (Fig. 4.6). All tasks with logic restraints that occur after the Clear Site task are now scheduled to start a day later. For example, in Figure 4.5 the Building Layout task was scheduled for Monday, August 22. In Figure 4.6, the updated start date for Building Layout is Tuesday, August 23.

Entering the Actual Task Duration

The Task Update dialog box allows you to enter the actual task duration as opposed to actual start and finish dates (Fig. 4.7).

PRACTICE THIS SKILL

1. Open Chapter4.mpp on the CD.
2. Open the Task Update dialog box.
3. A task duration of 2d has been entered in the Actual dur: field. The baseline plan allotted a 1-day duration (1d) for this task. By entering the updated task duration, you have recorded that the actual progress took twice as long as anticipated.
4. Click OK.

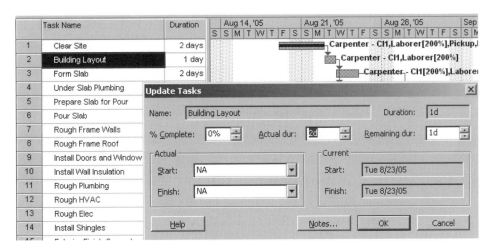

Figure 4.7 Building Layout Task (Before Update)

Notice the impact of the updated Building Layout task in Figure 4.8. Microsoft® Project automatically changed the following fields:

- **% Complete.** Changed to 100% complete because the actual finish date was entered, and the task is considered complete.
- **Remaining dur.** Changed to 0d because the actual finish date was entered, and the task is considered complete.

■ Notice that the Building Layout task bar on the Gantt Chart form in Figure 4.8 now has a dark solid line through it denoting actual progress. All tasks with logic restraints that occur after the Building Layout task are now starting a day later because of the day that was lost in completing Building Layout.

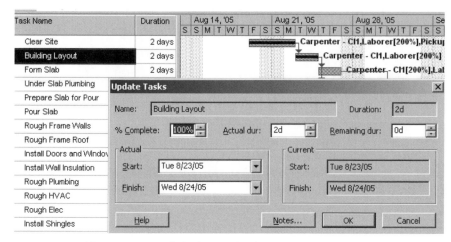

Figure 4.8 Building Layout Task (After Update)

Indicating Progress as a Percentage

You can also use the Task Update dialog box to enter the progress as a percentage (as opposed to actual task duration or actual start and finish dates).

PRACTICE THIS SKILL

1. Open Chapter4.mpp on the CD.
2. Open the Task Update Dialog Box for the Prepare Slab for Pour task.
3. Note that the % Complete: field contains a value of 66%.
4. Click OK.

Notice the impact of the updated Prepare Slab for Pour task in Figure 4.9. Microsoft® Project automatically changed the following fields:

■ **Actual dur.** Changed to 1.98d because almost 2 days of the baseline/original 3 days were expended.

■ **Remaining dur:** Changed to 1.02d because a little more than 1 day of the baseline/original 3 days is left to be expended.

■ **Actual Start:** Changed to Fri 8/26/05 to reflect the updated schedule.

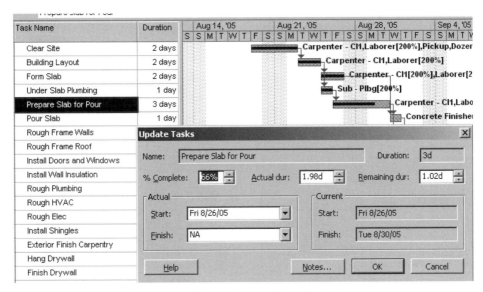

Figure 4.9 Update Tasks: Prepare Slab For Pour (After Update)

Selecting the Status Date

The status date represents the point at which progress is determined when updating the schedule. This status date is normally associated with progress payments and it is very important to control any changes that could affect the home's completion date.

1. Select the Project menu.
2. Click Project Information. The Project Information dialog box will appear (Figure 4.10).
3. Click the down arrow next to the Status date field to set a status date.

If you want the Status Date line to appear on the Gantt Chart, you'll need to perform the following steps:

1. Select the Tools menu.
2. Select Tracking.
3. Click Progress Lines. The Progress Lines dialog box will appear (Figure 4.11).
4. Check the Always display current progress line.
5. Check Baseline plan in the Display progress lines in relation to options box that is located in the lower right-hand portion of the Progress Lines dialog box. This will allow you to compare the current schedule to the baseline schedule. Notice that the 8/31/05 Status Date appears as a vertical line on the Gantt Chart (Fig. 4.12).

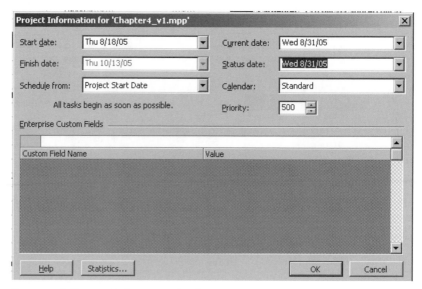

Figure 4.10   Project Information Dialog Box

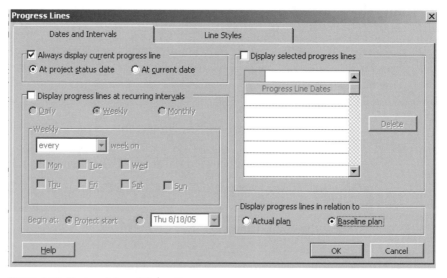

Figure 4.11   Project Lines Dialog Box

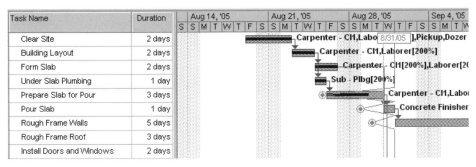

Figure 4.12 Updated Gantt Chart

Modifying Your Schedule

You must modify your schedule to show changes because it serves as your baseline home building plan. Many kinds of changes can occur over the duration of your project. You can change task durations if your crew size increases. You can also change the logic of task sequences and interrelationships to more accurately reflect interferences. You can add or delete tasks to reflect changes in the scope of the project, or you may need to add tasks for greater detail. You will definitely find areas to modify as the project proceeds and better data becomes available. Whatever the source, you must incorporate changes into the schedule for accurate updating and tracking. As a general rule, using the project name with the current date is the easiest way to organize and access updated schedules.

Now that you know how to monitor and update task durations, let's move on to updating task resources and costs.

CHAPTER 5

Updating Task Resources and Costs

Objectives

This chapter will teach you how to keep your schedule current by updating actual work progress and inputting resource and cost expenditures. Use the Chapter5.mpp file on the CD to follow along and perform the practice skills. After completing this chapter, you will be able to do the following:

- compare baseline costs to actual costs
- use updated cost tables
- record actual expenditures
- analyze costs
- split tasks

Updating Project Resources and Cost Information

Your ability to track and manage costs and keep your schedule current is directly related to the success of your project. Project costs are an accumulation of the resource costs of labor, equipment, materials and/or other resources. These costs are assigned to the task (allocated from the original estimate) when you develop the baseline schedule. By tracking actual costs and comparing these costs to the baseline budget, you can determine cost progress and earned value.

In order to control costs, you must determine whether the project is making or losing money according to the baseline plan. If there are problems, you must identify and address them. Home builders often use cost-loaded and cost-monitored schedules to communicate with the owner. This chapter shows you how to use Microsoft® Project to update costs to match the previously updated task durations and logic.

In order to keep your schedule current, you should continually track tasks and update their associated resources and costs. Microsoft® Project makes this process very easy for you to do. Once you enter the information, the program automatically updates the schedule. By using bar charts you can compare the baseline plan to the current progress and determine if any of your tasks are over or under budget.

Comparing Baseline Costs to Actual Costs
Current Schedule

The current schedule contains the schedule and logic modifications that you made to update the baseline schedule. It also includes the actual current data and dates, along with your actual resource expenditures. As you acquire actual cost information, you should enter it into the schedule to reflect the actual cost-to-date. Comparing the baseline costs to the actual costs of work performed enables you to forecast future cost based on the costs-to-date and your knowledge of the project.

Using Updated Cost Tables

In Chapter 4, we made task duration and logic changes. Now we must enter the actual cost information for the project. It is important to understand that Microsoft® Project creates new task cost totals based on duration changes. Therefore, the next thing we need to do is update the schedule with the actual cost expenditures and make some forecasts for future spending needs.

The easiest way to evaluate costs on-screen is by using the Cost Table. The Cost Table provides a convenient tool for tracking and analyzing costs. To access the Cost Table, simply click the View Menu, select Table: Entry, and click Cost (Fig. 5.1). The Cost Table will appear on your screen (Fig. 5.2).

Recording Actual Expenditures
Collecting Costs by Task

As you make physical progress on your project, you must record the actual cost expenditures and update your actual costs against the baseline estimated cost expenditures. You must collect costs by task in Microsoft® Project. Unfortunately, this is not the way most builders' cost accounting systems work. Costs are usually gathered by the cost account, rather than scheduled tasks. Labor time sheets, work measurement reports, purchase orders, and all other cost accounting documents are coded by cost account. The estimate is organized the same manner. Collecting costs by task means that you must add another step in your information gathering process.

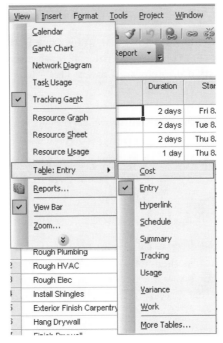

Figure 5.1 Accessing the Cost Table

Tracking Cost: Actual vs. Baseline Expenditures

In the following section, we will use the Cost Table to update individual tasks with current, actual cost information. First, let's update the Clear Site task.

	Task Name	Fixed Cost	Fixed Cost Accrual	Total Cost	Baseline	Variance	Actual	Remaining
1	Clear Site	$0.00	Prorated	$640.00	$640.00	$0.00	$640.00	$0.00
2	Building Layout	$0.00	Prorated	$448.00	$224.00	$224.00	$448.00	$0.00
3	Form Slab	$0.00	Prorated	$640.00	$640.00	$0.00	$640.00	$0.00
4	Under Slab Plumbing	$0.00	Prorated	$0.00	$0.00	$0.00	$0.00	$0.00
5	Prepare Slab for Pour	$0.00	Prorated	$672.00	$672.00	$0.00	$443.52	$228.48
6	Pour Slab	$0.00	Prorated	$384.00	$384.00	$0.00	$0.00	$384.00
7	Rough Frame Walls	$0.00	Prorated	$3,600.00	$3,600.00	$0.00	$0.00	$3,600.00
8	Rough Frame Roof	$0.00	Prorated	$2,160.00	$2,160.00	$0.00	$0.00	$2,160.00
9	Install Doors and Windows	$0.00	Prorated	$512.00	$512.00	$0.00	$0.00	$512.00
10	Install Wall Insulation	$0.00	Prorated	$768.00	$768.00	$0.00	$0.00	$768.00
11	Rough Plumbing	$0.00	Prorated	$0.00	$0.00	$0.00	$0.00	$0.00
12	Rough HVAC	$0.00	Prorated	$0.00	$0.00	$0.00	$0.00	$0.00

Figure 5.2 Cost Table

Clear Site Task

- **Baseline.** The Clear Site task in Figure 5.2 shows a Baseline cost of $640. Remember that the Baseline column represents the original plan for the cost required to complete this task. The Clear Site task is comprised of the following resources:

Resource	Hours	Cost
Carpenter	(8 hr/day) × 2 days = 16 hrs	16 × $12/hr = $192
Laborer (2)	(8 hr/day) × 2 days × 2 = 32 hrs	32 × $8/hr = $256
Pickup	(8 hr/day) × 2 days = 16 hrs	16 × $2/hr = $32
Dozer	(8 hr/day) × 2 days = 16 hrs	16 × $10/hr = $160
Baseline cost		$640

Use the Task Usage view to see all of the resources and costs associated with a specific task. To access this view, select the View menu, select Table, and click Usage. The Task Usage view will appear (Fig. 5.3). You should see that 80 hours are budgeted for the 4 resources. You can change this figure to reflect actual resource usage.

Task Name	Fixed Cost	Fixed Cost Accrual	Total Cost	Baseline	Variance	Actual	Details	Aug 21, '05 S	M
⊟ Clear Site	$80.00	Prorated	$720.00	$640.00	$80.00	$720.00	Work		40h
Carpenter - C/1			$192.00	$192.00	$0.00	$192.00	Work		8h
Laborer			$256.00	$256.00	$0.00	$256.00	Work		16h
Dozer			$160.00	$160.00	$0.00	$160.00	Work		8h
Pickup			$32.00	$32.00	$0.00	$32.00	Work		8h

Figure 5.3 Task Usage View

For example, let's change the dozer time from 16 hours to 15. (If you do not see the work hours, use the scroll bar on the lower right side of the screen until you get to August 21, 2005.) Click inside the Dozer Work field and change the value to 15 hours Notice that the actual cost information for the Clear Site task and the Laborer resource automatically changed to reflect the time difference. Close the file without saving your changes.

- **Total Cost.** The actual start and finish dates were changed to reflect actual progress for the Clear site task in Chapter 4. Because the actual duration remained at two days, Microsoft® Project did not change the value for this task in Total Cost column to reflect a change in cost (Figure 5.2).

- **Actual.** When the costs were originally tracked in the Actual cost column, there was a cost overrun on the task. Notice what happens when the Actual cost for the task is changed to $720 (Fig. 5.4):

o The Baseline remained the same.

o The Total Cost increased to $720.

o The Variance column now reflects the difference between the Total Cost minus the Baseline.

■ $720 (Total Cost) – $640 (Baseline) = $80 (Variance).

o The Fixed Cost increased by $80. Because the resource durations (Fig. 5.3) and costs remained the same, Microsoft® Project automatically placed the cost variance in the Fixed Cost column (Fig. 5.2).

o The Remaining column did not change because the activity is 100% complete (Fig. 5.2).

	Task Name	Fixed Cost	Fixed Cost Accrual	Total Cost	Baseline	Variance	Actual	Remaining
1	Clear Site	$80.00	Prorated	$720.00	$640.00	$80.00	$720.00	$0.00

Figure 5.4 Actual Cost Field

The Tracking Gantt provides another useful way to view cost information. Select the View menu and click Tracking Gantt. The Tracking Gantt view will appear on the screen (Fig. 5.5). In addition, the Gantt Chart at the right side of the screen shows the actual progress (solid blue bar) compared the baseline schedule (gray bar). Note that the Clear Site task is 100% complete and is one working day behind the baseline schedule.

	Task Name	Act. Start	Act. Finish	% Comp.	Phys. % Comp.	Act. Dur.	Rem. Dur.	Act. Cost	Act. Work
1	Clear Site	Fri 8/19/05	Mon 8/22/05	100%	0%	2 days	0 days	$720.00	80 hrs
2	Building Layout	Tue 8/23/05	Wed 8/24/05	100%	0%	2 days	0 days	$448.00	48 hrs
3	Form Slab	Thu 8/25/05	Fri 8/26/05	100%	0%	2 days	0 days	$640.00	64 hrs
4	Under Slab Plumbing	Thu 8/25/05	Thu 8/25/05	100%	0%	1 day	0 days	$0.00	16 hrs
5	Prepare Slab for Pour	Fri 8/26/05	NA	66%	0%	1.98 days	1.02 days	$443.52	47.52 hrs
6	Pour Slab	NA	NA	0%	0%	0 days	1 day	$0.00	0 hrs

Figure 5.5 Tracking Gantt View

Building Layout Task

The next task we'll update is Building Layout.

■ **Baseline.** The Building Layout task has a Baseline cost of $224 (Fig. 5.2). The Baseline cost plan for this task is as follows:

Resource	Hours	Cost
Carpenter	(8 hr/day) × 1 day = 8 hrs	8 × $12/hr = $96
Laborer (2)	(8 hr/day) × 1 day × 2 = 16 hrs	16 × $8/hr = $128
Baseline cost		$224

■ **Total Cost.** Note that the Total Cost column shows $448 (Fig. 5.2). When we updated this task in Chapter 4, we changed the Baseline duration from one day to an actual duration of two days. As a result, Microsoft® Project automatically doubled the requirements for all resources.

- **Actual.** When the costs were actually tracked in the field, only 30 hours of laborer time were expended (2 hours less than the projected 32 hours). We changed the Laborer resource from 32 hours to 30 using the Task Usage view (Fig. 5.6). Notice the impact of this change in Figure 5.7:

  o The Baseline cost remained the same ($224).

  o The Total Cost decreased to $432. This new total is $16 less than the amount shown in Figure 5.2 because we used less laborer time (30 hours as opposed to 32).

  o The Variance column reflects the difference between the Total Cost minus the Baseline.

    - $432 (Total Cost) – $224 (Baseline) = $208 (Variance).

  o The Fixed Cost column remained at $0 because the cost difference was changed as a Work type cost as opposed to a Fixed Cost (Figure 5.6).

  o The Remaining column did not change because the activity is 100% complete (Fig. 5.7).

	🛈	Task Name	Work	Duration	Start	Finish	Details	T	W
2	✓	⊟ Building Layout	48 hrs	2 days	Tue 8/23/05	Wed 8/24/05	Work	24h	24h
		Carpenter - C/1	16 hrs		Tue 8/23/05	Wed 8/24/05	Work	8h	8h
		Laborer	30 hrs		Tue 8/23/05	Wed 8/24/05	Work	16h	16h

Figure 5.6 Task Usage View

	Task Name	Fixed Cost	Fixed Cost Accrual	Total Cost	Baseline	Variance	Actual	Remaining
1	Clear Site	$80.00	Prorated	$720.00	$640.00	$80.00	$720.00	$0.00
2	Building Layout	$0.00	Prorated	$432.00	$224.00	$208.00	$432.00	$0.00

Figure 5.7 Building Layout (After Update)

Prepare Slab for Pour Task

The next task we'll update is Prepare Slab for Pour.

- **Baseline.** The Prepare Slab for Pour task has a Baseline cost of $672 (Fig. 5.2). The Baseline cost plan for this task is as follows:

Resource	Hours	Cost
Carpenter	(8 hr/day) × 3 days = 24 hrs	24 × $12/hr = $288
Laborer (2)	(8 hr/day) × 3 days × 2 = 48 hrs	48 × $8/hr = $384
Baseline cost		$672

- **Total Cost.** Note that the Total Cost ($672) is the same as the Baseline (Fig. 5.2).

- **Actual.** When the costs were tracked in the field, this task was 66% complete (Figure 5.8). Look at Figure 5.2 to evaluate a partially complete task:
  - o The Baseline column and the Total Cost column remained the same at $672.
  - o The Actual cost field reflects $443.52 (66% of $672).
  - o The Remaining column shows $228.38 (34% of $672).

	Task Name	Act. Start	Act. Finish	% Comp.	Phys. % Comp.	Act. Dur.	Rem. Dur.	Act. Cost	Act. Work	Aug 21, '05 S M T W T F S	Aug 28, '05 S M T W T F S
5	Prepare Slab for Pour	Fri 8/26/05	NA	66%	0%	1.98 days	1.02 days	$443.52	47.52 hrs		66%
6	Pour Slab	NA	NA	0%	0%	0 days	1 day	$0.00	0 hrs		0%

Figure 5.8 Prepare Slab For Pour (Before Update)

Analyzing Costs

Microsoft® Project provides a great way for you to analyze updated costs in your schedule. By using the Earned Value table, you can track your baseline cash flow against your actual progress, which allows you to ensure that you don't run out of money before the home is completed. The Earned Value table displays the actual percent complete of each task in terms of costs, so that you can forecast whether the task will finish over or under budget. By looking at all of the tasks at one time, you can forecast the cost of the house.

To view the Earned Value table, select the View menu, click Tables, and then click More Tables. The More Tables dialog box will appear (Fig. 5.9). Select Earned Value and click the Apply button.

Figure 5.9 More Table Dialog Box

Figure 5.10 shows the Earned Value table of the baseline and updated information as entered from Chapters 3 and 4. The explanation of the columns of the Earned Value table for the Clear Site task is as follows:

- **BCWS (Budgeted Cost of Work Scheduled).** Relates to the baseline (original) budget for the cumulative amount of work to be placed up to the status date. The baseline for the Clear Site task was $640. You can compare the BCWS field to the budgeted cost of work performed (BCWP) field to determine whether a task is behind or ahead of schedule in terms of cost.

- **BCWP.** Contains the cumulative value of the time-phased percent complete multiplied by the task's time-phased baseline cost through the status date of the schedule. This value is the earned value. The Clear Site is 100% complete. Therefore, 100% of the $640 for the Clear Site task should have been expended according to the baseline plan.

- **ACWP (Actual Cost of Work Performed).** Shows the costs incurred for the task through the status date of the schedule. There was an $80 overrun on the Clear Site task. Therefore, actual cost of this completed task was $720 ($640 + $80). You determine how and when the ACWP is calculated based on the settings you assigned in the Resource Information dialog box. The Resource Information dialog box settings are Standard Rate, Overtime Rate, Per Use Cost, and Cost Accrual, as well as the actual work reported through the status date.

- **SV (Schedule Variance).** Shows the earned value of a schedule variance through the status date of the schedule. The SV is the difference between BCWP and BCWS. The SV is $0 because the baseline and actual duration for the Clear Site task were both 2 days.

- **CV (Cost Variance).** Shows the earned value of a cost variance through the status date of the schedule. The CV is the difference between the BCWP and the ACWP. Because there was an $80 overrun on the Clear Site task, the CV column reflects a value of ($80).

- **EAC (Estimate at Completion).** Shows the total projected cost for a task, based on costs already incurred for the work performed in addition to costs planned for the task. The value in this field is $720 because the Clear Site task is 100% complete. However, if you had completed only 50% of the activity and had incurred a cost different than the 50% of the activity ($350) then the estimated cost at completion would have been modified to show the actual expenditure as it relates to physical progress.

- **BAC (Budgeted at Completion).** Shows the estimated cost of the activity at completion. For example: if you estimate that an activity will require 40 hours of work, with a wage of $10/hours, $300 worth of materials, and negligible equipment cost, then the budget cost at completion will be 40 hrs × $10/hr + $300 + $0 = $700

- **VAC (Variance at Completion).** Shows the difference between the BAC and the EAC. The value in this field is $80 (the total amount of the cost overrun) because the Clear Site task is 100% complete.

	Task Name	BCWS	BCWP	ACWP	SV	CV	EAC	BAC
1	Clear Site	$640.00	$640.00	$720.00	$0.00	($80.00)	$720.00	$640.00
2	Building Layout	$224.00	$224.00	$432.00	$0.00	($208.00)	$432.00	$224.00
3	Form Slab	$640.00	$640.00	$640.00	$0.00	$0.00	$640.00	$640.00
4	Under Slab Plumbing	$0.00	$0.00	$0.00	$0.00	$0.00	$0.00	$0.00
5	Prepare Slab for Pour	$672.00	$443.80	$443.52	($228.20)	$0.28	$671.58	$672.00

Figure 5.10 Earned Value Table

Entering Split Tasks

Situations can arise during the home building process in which certain tasks must be delayed because the necessary resources may become temporarily unavailable. In this case, you must reschedule the remaining work on that task for a later time. Microsoft® Project allows you to split tasks to allow the task to proceed until the resource leaves and resume when the resource returns. It is best to plan splits in advance. Microsoft® Project will automatically update the partially completed task's actual and scheduled work as you create the split.

To split a task, click the Gantt Chart icon on the View bar, and click the appropriate task. In Figure 5.11, the Building Layout task is selected. When you click the Split Task icon on the Standard toolbar the Split Task dialog box will appear (Fig. 5.12). Point to the bar of the task you want to split on the Gantt Chart. Click and drag the split bar to the right until the appropriate finish split date appears. Then select the View menu and click Gantt Chart to see the results of the split task (Fig. 5.13). Microsoft® Project will reallocate all resources and costs to the broken segments.

PRACTICE THIS SKILL

1. Open Chapter5.mpp on the CD.

2. Split the Form Slab task.

3. View the results.

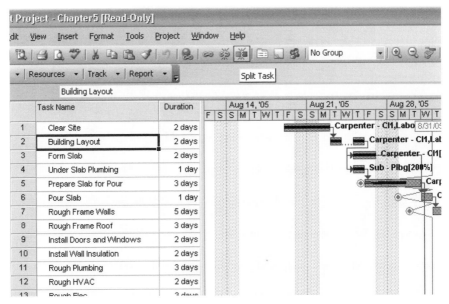

Figure 5.11 Building Layout Task

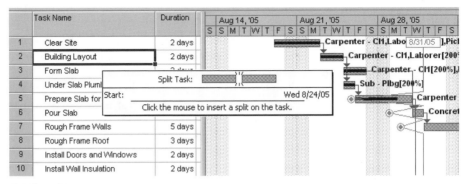

Figure 5.12 Split Task Dialog Box

	Task Name	Duration
1	Clear Site	2 days
2	Building Layout	2 days
3	Form Slab	2 days
4	Under Slab Plumbing	1 day
5	Prepare Slab for Pour	3 days
6	Pour Slab	1 day
7	Rough Frame Walls	5 days

Figure 5.13 Split Tasks (Results)

Now that you know how to update task resources and costs, let's move on to the final skill—printing reports.

CHAPTER 6

Reports

Objectives

This chapter will teach you how to access and evaluate project data and print reports for distribution. Use the Chapter6.mpp file on the CD to follow along and perform the practice skills. After completing this chapter, you will be able to do the following:

- execute prints and copies
- print views
- print reports
- edit reports
- copy reports

Good Communications are a Necessity

Good scheduling involves distributing information and receiving feedback on all aspects of your plan. Developing a schedule at the beginning of your home building project forces you to plan, organize, sequence, and show the interrelationships among different tasks of the project. However, this information is useless unless you communicate it to all key parties involved in the project. Microsoft® Project allows you to print graphical and tabular views and reports to meet a variety of needs. A graphical report is a pictorial printout or plot representation of the project status with respect to costs, resources, or progress. A tabular report displays information in columns or in spreadsheet format for easy analysis and comparison.

Executing Prints and Copies

The printing process can be as simple as reviewing a file prior to printing it or as involved as creating and custom reports. To preview the document on-screen before you actually print it, select the File menu, and click Print Preview. This feature allows you to fine-tune the file. It can save you a lot of money on wasted paper. When you are ready to actually print the file, select Print from the File menu.

To produce preconfigured or customized reports, select the View menu, and click Reports.

Select the report style that best meets your needs. You can use the template reports or create customized reports. You can enhance the visual appeal of your reports by adding header and/or footers that include your company logo and the project name. You can also scale information to fit on a single page.

You can print a range of pages (defined by page numbers or dates), suppress (not print) blank pages, and print multiple copies to make printing as efficient as possible. If you are running other Microsoft programs on your computer, Microsoft® Project will use the same printer by default. If not, you must select a printer before you can print.

Print Views—File Menu

File Menu

When printing a view, by default, the number of columns displayed on the screen determines the number of columns that will be printed. For example, if you display the first five columns in the Gantt Chart view (ID, Indicators, Task Name, Duration, and Start), those five columns will appear on the printed pages. On most views, you can specify the exact number of columns you want to print. Views that are too large to print on a single page are printed down and across, left to right, starting in the upper-left corner of the view. The pages will be numbered accordingly. Microsoft® Project allows you to move within a view to display additional project information and adjust the timescale in a chart view to display additional graphical information. You can use the Zoom In and Zoom Out buttons on the Standard toolbar to adjust these timescales. Or, you can use the Timescale dialog box, which you can select from the Format menu.

Using different views is the easiest way to diversify the formatting and display of on-screen information. You can easily display Microsoft® Project's preconfigured views by using the View menu or the View bar located on the left side of the program window (Fig. 6.1).

The on-screen view options are as follows:

- **Calendar.** Allows you to view the project in a monthly format (Fig. 6.2).
- **Gantt Chart.** Provides an easy-to-use graphical representation of the schedule. Commonly used by home builders, this view was used for most of the figures in this book (Fig. 6.3).
- **Network Diagram.** Provides a clear picture of the logic flow for task relationships (Fig. 6.4).
- **Task Usage.** Allows you to evaluate daily task resource requirements (Fig. 6.5).
- **Tracking Gantt.** Provides valuable task tabular progress information along with the graphical progress of the task (Fig. 6.6).

Figure 6.1 View Menu

- **Resource Graph.** Graphically shows the resource requirements by time units (Fig. 6.7).
- **Resource Sheet.** Depicts the resource list as produced for the schedule (Fig. 6.8).
- **Resource Usage.** Shows the total usage for each resource by task (Fig. 6.9).
- **More Views.** When this option is selected, the More Views dialog box will appear (Fig. 6.10). You can change views by selecting any of the preconfigured views and then clicking Apply. When you select Print or Print Preview from the File menu, the new view will appear. To change views, simply click the desired view option from the View menu or the View bar (Fig. 6.1).

Print Reports—View Menu

Microsoft® Project includes 30 preconfigured report types which you can easily modify to meet your unique needs. To view the overall report categories, select the View menu and click Reports. The Reports dialog box will appear (Figure 6.11). The following six report categories will be displayed:

- overview
- current activities
- costs

September 2005

Sunday	Monday	Tuesday	Wednesday	Thursday	Friday	Saturday
					1 Rough Frame Walls, 5 days; Prepare Driveways and Walks, 4 days	2	3
4	5 Rough Frame Walls, 5 days; Prepare Driveways and Walks, 4 days	6 Pour Driveways and Walks,	7	8 Rough Frame Roof, 3 days; Install Doors and Windows, 2 days	9	10
11 Rough Frame Roof, 3 days	12	13 Rough Plumbing, 3 days; Rough HVAC, 2 days; Rough Elec, 3 days; Install Shingles, 2 days	14	15 Install Wall Insulation, 2 days	16	17
18 Install Wall Insulation, 2 days	19	20 Exterior Finish Carpentry, 4 days; Hang Drywall, 3 days	21	22	23 Finish Drywall, 3 days	24
25 Finish Drywall, 3 days	26	27 Place Exterior Brick, 4 days; Place Cabinets, 1 day; Interior Finish Carpentry, 3 days	28	29	30 Exterior Paint, 3 days	

Figure 6.2 Calendar View

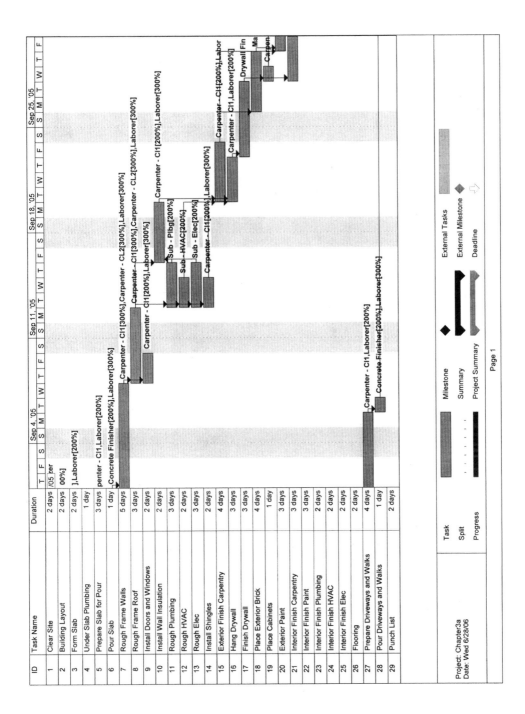

Figure 6.3 Gantt Chart View

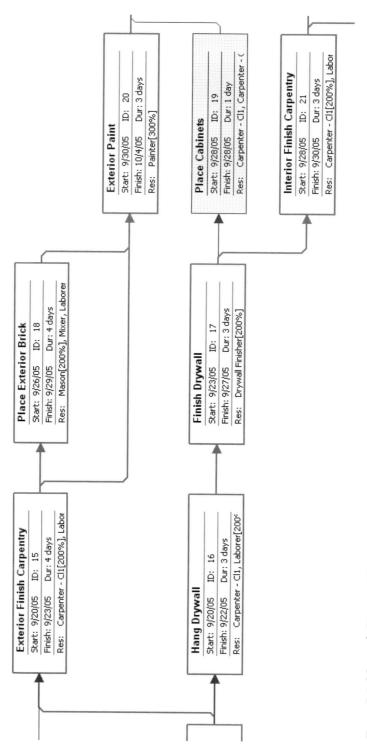

Figure 6.4 Network Diagram View

Chapter3a

ID	Task Name	Fixed Cost	Fixed Cost Accrual	Total Cost	Baseline	Variance	Actual	Details	T	F	S	S	M
1	Clear Site	$80.00	Prorated	$720.00	$640.00	$80.00	$720.00	Work					
	Carpenter - C/1			$192.00	$192.00	$0.00	$192.00	Work					
	Laborer			$256.00	$256.00	$0.00	$256.00	Work					
	Dozer			$160.00	$160.00	$0.00	$160.00	Work					
	Pickup			$32.00	$32.00	$0.00	$32.00	Work					
2	Building Layout	$0.00	Prorated	$432.00	$224.00	$208.00	$432.00	Work					
	Carpenter - C/1			$192.00	$96.00	$96.00	$192.00	Work					
	Laborer			$240.00	$128.00	$112.00	$240.00	Work					
3	Form Slab	$0.00	Prorated	$640.00	$640.00	$0.00	$640.00	Work					
	Carpenter - C/1			$384.00	$384.00	$0.00	$384.00	Work					
	Laborer			$256.00	$256.00	$0.00	$256.00	Work					
4	Under Slab Plumbing	$0.00	Prorated	$0.00	$0.00	$0.00	$0.00	Work					
	Sub - Plbg			$0.00	$0.00	$0.00	$0.00	Work					
5	Prepare Slab for Pour	$0.00	Prorated	$672.00	$672.00	$0.00	$443.52	Work					
	Carpenter - C/1			$288.00	$288.00	$0.00	$190.08	Work					
	Laborer			$384.00	$384.00	$0.00	$253.44	Work					
6	Pour Slab	$0.00	Prorated	$384.00	$384.00	$0.00	$0.00	Work					
	Laborer			$192.00	$192.00	$0.00	$0.00	Work					
	Concrete Finisher			$192.00	$192.00	$0.00	$0.00	Work					
7	Rough Frame Walls	$0.00	Prorated	$3,600.00	$3,600.00	$0.00	$0.00	Work	72h	72h			72h
	Carpenter - C/1			$1,440.00	$1,440.00	$0.00	$0.00	Work	24h	24h			24h
	Carpenter - CL2			$1,200.00	$1,200.00	$0.00	$0.00	Work	24h	24h			24h
	Laborer			$960.00	$960.00	$0.00	$0.00	Work	24h	24h			24h
8	Rough Frame Roof	$0.00	Prorated	$2,160.00	$2,160.00	$0.00	$0.00	Work					
	Carpenter - C/1			$864.00	$864.00	$0.00	$0.00	Work					
	Carpenter - CL2			$720.00	$720.00	$0.00	$0.00	Work					
	Laborer			$576.00	$576.00	$0.00	$0.00	Work					
9	Install Doors and Windows	$0.00	Prorated	$512.00	$512.00	$0.00	$0.00	Work					
	Carpenter - C/1			$384.00	$384.00	$0.00	$0.00	Work					
	Laborer			$128.00	$128.00	$0.00	$0.00	Work					
10	Install Wall Insulation	$0.00	Prorated	$768.00	$768.00	$0.00	$0.00	Work					
	Carpenter - C/1			$384.00	$384.00	$0.00	$0.00	Work					
	Laborer			$384.00	$384.00	$0.00	$0.00	Work					
11	Rough Plumbing	$0.00	Prorated	$0.00	$0.00	$0.00	$0.00	Work					
	Sub - Plbg			$0.00	$0.00	$0.00	$0.00	Work					
12	Rough HVAC	$0.00	Prorated	$0.00	$0.00	$0.00	$0.00	Work					
	Sub - HVAC			$0.00	$0.00	$0.00	$0.00	Work					
13	Rough Elec	$0.00	Prorated	$0.00	$0.00	$0.00	$0.00	Work					
	Sub - Elec			$0.00	$0.00	$0.00	$0.00	Work					
14	Install Shingles	$0.00	Prorated	$768.00	$768.00	$0.00	$0.00	Work					
	Carpenter - C/1			$384.00	$384.00	$0.00	$0.00	Work					
	Laborer			$384.00	$384.00	$0.00	$0.00	Work					
15	Exterior Finish Carpentry	$0.00	Prorated	$1,024.00	$1,024.00	$0.00	$0.00	Work					
	Carpenter - C/1			$768.00	$768.00	$0.00	$0.00	Work					
	Laborer			$256.00	$256.00	$0.00	$0.00	Work					
16	Hang Drywall	$0.00	Prorated	$672.00	$672.00	$0.00	$0.00	Work					
	Carpenter - C/1			$288.00	$288.00	$0.00	$0.00	Work					
	Laborer			$384.00	$384.00	$0.00	$0.00	Work					
17	Finish Drywall	$0.00	Prorated	$432.00	$432.00	$0.00	$0.00	Work					
	Drywall Finisher			$432.00	$432.00	$0.00	$0.00	Work					
18	Place Exterior Brick	$0.00	Prorated	$1,664.00	$1,664.00	$0.00	$0.00	Work					
	Laborer			$512.00	$512.00	$0.00	$0.00	Work					
	Mason			$896.00	$896.00	$0.00	$0.00	Work					

Sep 4, '05

Figure 6.5 Task Usage View

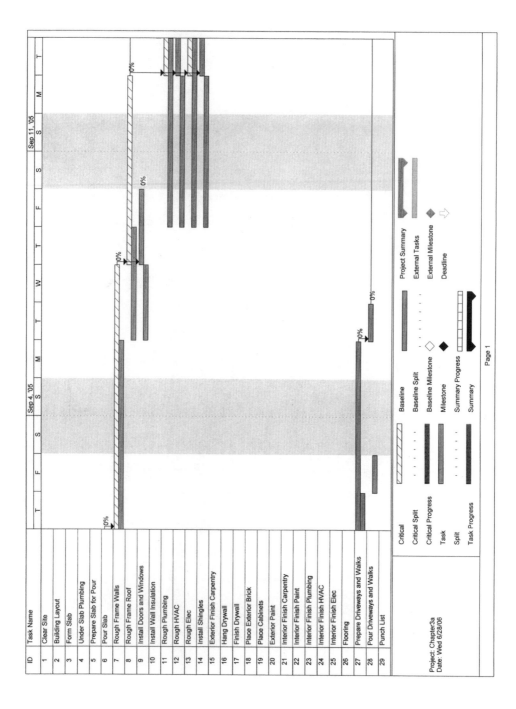

Figure 6.6 Tracking Gantt View

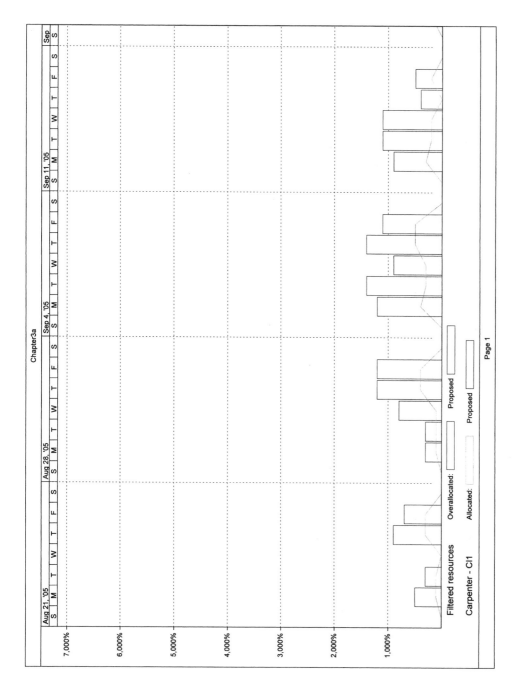

Figure 6.7 Resource Graph Usage View

ID	●	Resource Name	Type	Material Label	Group	Max. Units	Std. Rate	Ovt. Rate	Cost/Use	Accrue At
1	▦	Carpenter - Cl1	Work			600%	$12.00/hr	$18.00/hr	$0.00	Prorated
2		Carpenter - CL2	Work			400%	$10.00/hr	$15.00/hr	$0.00	Prorated
3		Drywall Finisher	Work			400%	$9.00/hr	$13.50/hr	$0.00	Prorated
4		Laborer	Work			600%	$8.00/hr	$12.00/hr	$0.00	Prorated
5		Mason	Work		Masonry Crew	400%	$14.00/hr	$21.00/hr	$0.00	Prorated
6		Mixer	Work		Masonry Crew	400%	$8.00/hr	$12.00/hr	$0.00	Prorated
7		Concrete Finisher	Work			400%	$12.00/hr	$18.00/hr	$0.00	Prorated
8		Painter	Work			500%	$9.00/hr	$13.50/hr	$0.00	Prorated
9		Superintendent	Work			100%	$16.00/hr	$24.00/hr	$0.00	Prorated
10		Sub - Elec	Work			500%	$0.00/hr	$0.00/hr	$0.00	Prorated
11		Sub - Flooring	Work			500%	$0.00/hr	$0.00/hr	$0.00	Prorated
12		Sub - HVAC	Work			500%	$0.00/hr	$0.00/hr	$0.00	Prorated
13		Sub - Plbg	Work			500%	$0.00/hr	$0.00/hr	$0.00	Prorated
14		Dozer	Work			100%	$10.00/hr	$0.00/hr	$0.00	Prorated
15		Trailer	Work			100%	$8.00/hr	$0.00/hr	$0.00	Prorated
16		Masonry Saw	Work		Masonry Crew	100%	$1.00/hr	$0.00/hr	$0.00	Prorated
17		Radial Arm Saw	Work			100%	$1.00/hr	$0.00/hr	$0.00	Prorated
18		Generator	Work			100%	$2.00/hr	$0.00/hr	$0.00	Prorated
19		Pickup	Work			100%	$2.00/hr	$0.00/hr	$0.00	Prorated
20		Mortar Mixer	Work		Masonry Crew	100%	$2.00/hr	$0.00/hr	$20.00	Prorated
21		Concrete	Material	CY		100%	$50.00		$0.00	Prorated

Figure 6.8 Resource Sheet View

ID	ⓘ	Resource Name	Work	Details	Aug 21, '05 S	M	T	W	T	F	S	Aug 28, '05 S	M	T	W	T	F
1	📅	**Carpenter - Cl1**	568 hrs	Work		8h	8h	8h	8h	24h			8h	8h	8h	32h	32h
	🔔	Clear Site	16 hrs	Work		8h	8h										
		Building Layout	16 hrs	Work		8h	8h										
		Form Slab	32 hrs	Work													
		Prepare Slab for I	24 hrs	Work			8h	0h	8h	8h							
		Rough Frame Wa	120 hrs	Work				0h	16h	16h			8h	8h	8h		24h
		Rough Frame Rou	72 hrs	Work						8h						24h	24h
		Install Doors and	32 hrs	Work													
		Install Wall Insula	32 hrs	Work													
		Install Shingles	32 hrs	Work													
		Exterior Finish Ca	64 hrs	Work													
		Hang Drywall	24 hrs	Work													
		Place Cabinets	8 hrs	Work											8h	8h	8h
		Interior Finish Cai	48 hrs	Work													
		Prepare Driveway	32 hrs	Work													
		Punch List	16 hrs	Work													
2		**Carpenter - CL2**	200 hrs	Work												24h	24h
		Rough Frame Wa	120 hrs	Work												24h	24h
		Rough Frame Rou	72 hrs	Work													
		Place Cabinets	8 hrs	Work													
3		**Drywall Finisher**	48 hrs	Work													
		Finish Drywall	48 hrs	Work													
4	🔔	**Laborer**	726 hrs	Work		16h	16h	0h	30h	32h			16h	16h	40h	40h	40h
		Clear Site	32 hrs	Work		16h	16h										
		Building Layout	30 hrs	Work													
		Form Slab	32 hrs	Work													
		Prepare Slab for I	48 hrs	Work			16h	0h	14h	16h			16h	16h	24h		24h
		Pour Slab	24 hrs	Work				0h	16h	16h			16h	16h			
		Rough Frame Wa	120 hrs	Work						16h							
		Rough Frame Rou	72 hrs	Work													
		Install Doors and	16 hrs	Work													
		Install Wall Insula	48 hrs	Work													
		Install Shingles	48 hrs	Work													
		Exterior Finish Ca	32 hrs	Work													
		Hang Drywall	48 hrs	Work													
		Place Exterior Bri	64 hrs	Work													
		Interior Finish Cai	24 hrs	Work													
		Prepare Driveway	64 hrs	Work													
		Pour Driveways a	24 hrs	Work													
5		**Mason**	64 hrs	Work													
		Place Exterior Bri	64 hrs	Work													
6		**Mixer**	32 hrs	Work													
		Place Exterior Bri	32 hrs	Work													
7		**Concrete Finisher**	32 hrs	Work													
		Pour Slab	16 hrs	Work										16h	16h		16h
		Pour Driveways a	16 hrs	Work													16h
8		**Painter**	160 hrs	Work													
		Exterior Paint	72 hrs	Work													
		Interior Finish Pai	72 hrs	Work													
		Punch List	16 hrs	Work													
9		**Superintendent**	0 hrs	Work													
10		**Sub - Elec**	96 hrs	Work													
		Rough Elec	48 hrs	Work													

Figure 6.9 Resource Usage View

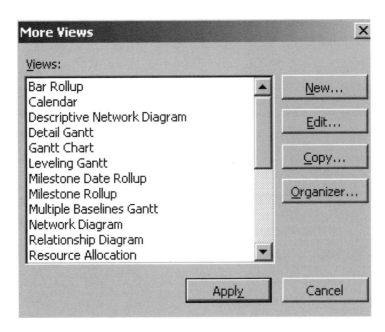

Figure 6.10 More Views Dialog Box

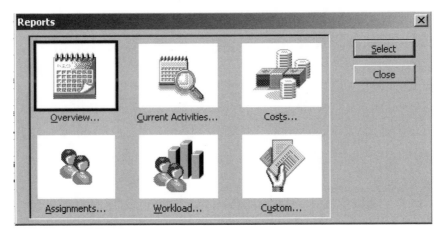

Figure 6.11 Reports Dialog Box

- assignments
- workload
- custom

Overview

Overview reports display information over the duration of the project. You can print five different reports:

- Project Summary
- Top-Level Tasks
- Critical Tasks
- Milestones
- Working Days

To access the Overview reports, select the Overview button from the Reports dialog box, and click Select. The Overview Reports dialog box will appear (Fig. 6.12). Select the type of report that best suits your needs and click Print. See Figure 6.13 for a sample Critical Tasks report.

Current Activities

Current activity reports display information about the status of certain tasks. You can print six different reports:

- Unstarted Tasks
- Tasks Starting Soon
- Tasks In Progress
- Completed Tasks

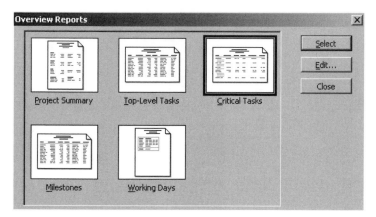

Figure 6.12 Overview Reports Dialog Box

ID	Task Name	Duration			Start	Finish	Predecessors	Resource Names
5	Prepare Slab for Pour	3 days			Fri 8/26/05	Tue 8/30/05	4	Carpenter - CI1,Laborer[20C
	ID *Successor Name*	*Type*	*Lag*					
	6 *Pour Slab*	*FS*	*0 days*					
	27 *Prepare Driveways and Walks*	*FS*	*0 days*					
6	Pour Slab	1 day			Wed 8/31/05	Wed 8/31/05	3,5	Concrete Finisher[200%],La
	ID *Successor Name*	*Type*	*Lag*					
	7 *Rough Frame Walls*	*FS*	*0 days*					
7	Rough Frame Walls	5 days			Thu 9/1/05	Wed 9/7/05	6	Carpenter - CI1[300%],Carp
	ID *Successor Name*	*Type*	*Lag*					
	8 *Rough Frame Roof*	*FS*	*0 days*					
	9 *Install Doors and Windows*	*FS*	*0 days*					
8	Rough Frame Roof	3 days			Thu 9/8/05	Mon 9/12/05	7	Carpenter - CI1[300%],Carp
	ID *Successor Name*	*Type*	*Lag*					
	10 *Install Wall Insulation*	*FS*	*0 days*					
	11 *Rough Plumbing*	*FS*	*0 days*					
	12 *Rough HVAC*	*FS*	*0 days*					
	13 *Rough Elec*	*FS*	*0 days*					
	14 *Install Shingles*	*FS*	*0 days*					
10	Install Wall Insulation	2 days			Fri 9/16/05	Mon 9/19/05	8,11,13	Carpenter - CI1[200%],Labo
	ID *Successor Name*	*Type*	*Lag*					
	15 *Exterior Finish Carpentry*	*FS*	*0 days*					
	16 *Hang Drywall*	*FS*	*0 days*					
11	Rough Plumbing	3 days			Tue 9/13/05	Thu 9/15/05	8	Sub - Plbg[200%]
	ID *Successor Name*	*Type*	*Lag*					
	10 *Install Wall Insulation*	*FS*	*0 days*					
13	Rough Elec	3 days			Tue 9/13/05	Thu 9/15/05	8	Sub - Elec[200%]
	ID *Successor Name*	*Type*	*Lag*					
	10 *Install Wall Insulation*	*FS*	*0 days*					
15	Exterior Finish Carpentry	4 days			Tue 9/20/05	Fri 9/23/05	10,14	Carpenter - CI1[200%],Labo
	ID *Successor Name*	*Type*	*Lag*					
	18 *Place Exterior Brick*	*FS*	*0 days*					
	20 *Exterior Paint*	*FS*	*0 days*					
16	Hang Drywall	3 days			Tue 9/20/05	Thu 9/22/05	10,12	Carpenter - CI1,Laborer[20C
	ID *Successor Name*	*Type*	*Lag*					
	17 *Finish Drywall*	*FS*	*0 days*					
17	Finish Drywall	3 days			Fri 9/23/05	Tue 9/27/05	16	Drywall Finisher[200%]
	ID *Successor Name*	*Type*	*Lag*					
	19 *Place Cabinets*	*FS*	*0 days*					
	21 *Interior Finish Carpentry*	*FS*	*0 days*					
18	Place Exterior Brick	4 days			Mon 9/26/05	Thu 9/29/05	15	Mason[200%],Mixer,Laborei
	ID *Successor Name*	*Type*	*Lag*					
	20 *Exterior Paint*	*FS*	*0 days*					

Figure 6.13 Critical Tasks Report

- Should Have Started Tasks
- Slipping Tasks

To access the Current Activities reports, select Current Activities from the Reports dialog box and click Select. The Current Activities dialog box will appear (Fig. 6.14). Select the type of report that best suits your needs and click Print. See Figure 6.15 for a sample Tasks in Progress report.

Costs

Cost reports display financial information about your project. You can print five different reports:

- Cash Flow
- Budget
- Overbudget Tasks
- Overbudget Resources
- Earned Value

To access the Costs reports, select Costs from the Reports dialog box and click Select.

The Cost Reports dialog box will appear (Fig. 6.16). Select the type of report that best suits your needs and click Print. See Figure 6.17 for a sample Cash Flow report.

Assignments

Assignment reports display task resource information. You can print four different reports:

- Who Does What
- Who Does What When

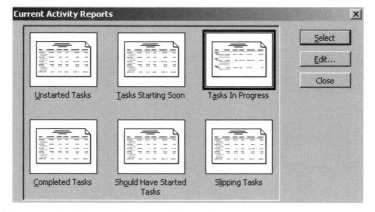

Figure 6.14 Current Activities Dialog Box

Tasks In Progress as of Wed 6/28/06
Chapter3a

ID	Task Name	Duration	Start	Finish	Predecessors	Resource Names
August 2005						
5	Prepare Slab for Pour	3 days	Fri 8/26/05	Tue 8/30/05	4	Carpenter - CI1,Laborer[20C

ID	Resource Name	Units	Work	Delay	Start	Finish
1	Carpenter - CI1	100%	24 hrs	0 days	Fri 8/26/05	Tue 8/30/05
4	Laborer	200%	48 hrs	0 days	Fri 8/26/05	Tue 8/30/05

Figure 6.15 Tasks in Progress Report

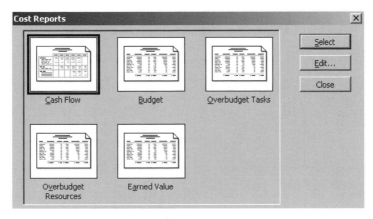

Figure 6.16 Cost Reports Dialog Box

- To-Do List
- Overallocated Resources

To access the Assignment reports, select Assignments from the Reports dialog box and click Select. The Assignment Reports dialog box will appear (Fig. 6.18). Select the type of report that best suits your needs and click Print. See Figure 6.19 for a sample Who Does What report.

Workload

Workload reports display task usage information. You can print two different reports:

- Task Usage
- Resource Usage

To access the Workload reports, select Workload from the Reports dialog box and click Select. The Workload Reports dialog box will appear (Fig. 6.20). Select the type of report that best suits your needs and click Print. See Figure 6.21 for a sample Resource Usage report.

Custom

You can print all of the previously mentioned reports, or if they don't meet your needs there are eight additional Custom reports that you can modify to meet your specific needs:

- Base Calendar
- Crosstab
- Resource
- Resource (Material)

	8/14/05	8/21/05	8/28/05	9/4/05	9/11/05	9/18/05	9/25/05	10/2/05	10/9/05
Clear Site	$360.00								
Building Layout		$360.00							
Form Slab		$432.00							
Under Slab Plumbing		$640.00							
Prepare Slab for Pour		$224.00	$448.00						
Pour Slab			$384.00						
Rough Frame Walls			$1,440.00	$2,160.00					
Rough Frame Roof				$1,440.00	$720.00				
Install Doors and Windows				$512.00					
Rough Wall Insulation					$384.00				
Rough Plumbing						$384.00			
Rough HVAC									
Rough Elec									
Install Shingles					$768.00				
Exterior Finish Carpentry						$1,024.00			
Hang Drywall						$672.00			
Finish Drywall						$144.00	$288.00		
Place Exterior Brick							$1,664.00		
Place Cabinets							$176.00		
Exterior Paint							$716.00		
Interior Finish Carpentry							$768.00	$1,432.00	
Interior Finish Paint								$648.00	
Interior Finish Plumbing									
Interior Finish HVAC									
Interior Finish Elec									
Flooring									
Prepare Driveways and Walks			$672.00	$224.00					
Pour Driveways and Walks				$384.00					
Punch List									$336.00
Total	$360.00	$1,656.00	$2,944.00	$4,720.00	$1,872.00	$2,224.00	$3,612.00	$2,080.00	$336.00

Figure 6.17 Cash Flow Report

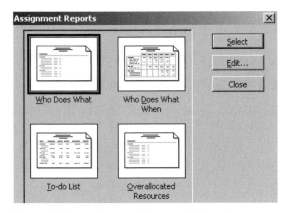

Figure 6.18 Assignment Reports Dialog Box

■ Resource (Work)
■ Resource Usage (Material)
■ Resource Usage (Work)
■ Task

To access Custom reports, select Custom from the Reports dialog box and click Select. The Custom Reports dialog box will appear (Figure 6.22). Select the type of report that best suits your needs and click Print. We will cover Custom Reports in the next section.

PRACTICE THIS SKILL

1. Open the Chapter6.mpp file.
2. Run the following reports:
   a. Critical Tasks
   b. Tasks in Progress
   c. Cash Flow
   d. Who Does What
   e. Resource Usage
3. Print the reports.

Editing Custom Reports

The ability to edit these reports to meet your unique needs is a valuable tool. To edit a report, use the Edit button from any of the six reports categories of the Reports dialog box (Figure 6.11).

ID	ⓘ	Resource Name	Work

1 — Carpenter - CI1 — 568 hrs

ID	Task Name	Units	Work	Delay	Start	Finish
7	Rough Frame Walls	300%	120 hrs	0 days	Thu 9/1/05	Wed 9/7/05
9	Install Doors and Windows	200%	32 hrs	0 days	Thu 9/8/05	Fri 9/9/05
15	Exterior Finish Carpentry	200%	64 hrs	0 days	Tue 9/20/05	Fri 9/23/05
21	Interior Finish Carpentry	200%	48 hrs	0 days	Wed 9/28/05	Fri 9/30/05
1	Clear Site	100%	16 hrs	0 days	Fri 8/19/05	Mon 8/22/05
2	Building Layout	100%	16 hrs	0 days	Tue 8/23/05	Thu 8/25/05
3	Form Slab	200%	32 hrs	0 days	Thu 8/25/05	Fri 8/26/05
5	Prepare Slab for Pour	100%	24 hrs	0 days	Fri 8/26/05	Tue 8/30/05
8	Rough Frame Roof	300%	72 hrs	0 days	Thu 9/8/05	Mon 9/12/05
10	Install Wall Insulation	200%	32 hrs	0 days	Fri 9/16/05	Mon 9/19/05
14	Install Shingles	200%	32 hrs	0 days	Tue 9/13/05	Wed 9/14/05
16	Hang Drywall	100%	24 hrs	0 days	Tue 9/20/05	Thu 9/22/05
19	Place Cabinets	100%	8 hrs	0 days	Wed 9/28/05	Wed 9/28/05
27	Prepare Driveways and Walks	100%	32 hrs	0 days	Wed 8/31/05	Mon 9/5/05
29	Punch List	100%	16 hrs	0 days	Wed 10/12/05	Thu 10/13/05

2 — Carpenter - CL2 — 200 hrs

ID	Task Name	Units	Work	Delay	Start	Finish
7	Rough Frame Walls	300%	120 hrs	0 days	Thu 9/1/05	Wed 9/7/05
8	Rough Frame Roof	300%	72 hrs	0 days	Thu 9/8/05	Mon 9/12/05
19	Place Cabinets	100%	8 hrs	0 days	Wed 9/28/05	Wed 9/28/05

3 — Drywall Finisher — 48 hrs

ID	Task Name	Units	Work	Delay	Start	Finish
17	Finish Drywall	200%	48 hrs	0 days	Fri 9/23/05	Tue 9/27/05

4 — Laborer — 726 hrs

ID	Task Name	Units	Work	Delay	Start	Finish
7	Rough Frame Walls	300%	120 hrs	0 days	Thu 9/1/05	Wed 9/7/05
9	Install Doors and Windows	300%	16 hrs	0 days	Thu 9/8/05	Thu 9/8/05
15	Exterior Finish Carpentry	300%	32 hrs	0 days	Tue 9/20/05	Wed 9/21/05
21	Interior Finish Carpentry	300%	24 hrs	0 days	Wed 9/28/05	Wed 9/28/05
1	Clear Site	200%	32 hrs	0 days	Fri 8/19/05	Mon 8/22/05
2	Building Layout	200%	30 hrs	0 days	Tue 8/23/05	Thu 8/25/05
3	Form Slab	200%	32 hrs	0 days	Thu 8/25/05	Fri 8/26/05
5	Prepare Slab for Pour	200%	48 hrs	0 days	Fri 8/26/05	Tue 8/30/05
8	Rough Frame Roof	300%	72 hrs	0 days	Thu 9/8/05	Mon 9/12/05
10	Install Wall Insulation	300%	48 hrs	0 days	Fri 9/16/05	Mon 9/19/05
14	Install Shingles	300%	48 hrs	0 days	Tue 9/13/05	Wed 9/14/05
16	Hang Drywall	200%	48 hrs	0 days	Tue 9/20/05	Thu 9/22/05
18	Place Exterior Brick	200%	64 hrs	0 days	Mon 9/26/05	Thu 9/29/05
6	Pour Slab	300%	24 hrs	0 days	Wed 8/31/05	Wed 8/31/05
27	Prepare Driveways and Walks	200%	64 hrs	0 days	Wed 8/31/05	Mon 9/5/05
28	Pour Driveways and Walks	300%	24 hrs	0 days	Tue 9/6/05	Tue 9/6/05

5 — Mason — 64 hrs

ID	Task Name	Units	Work	Delay	Start	Finish
18	Place Exterior Brick	200%	64 hrs	0 days	Mon 9/26/05	Thu 9/29/05

Figure 6.19 Who Does What Report

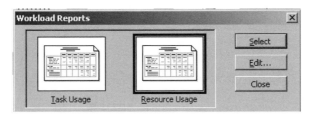

Figure 6.20 Workload Reports Dialog Box

For example, if you select the Critical Task report from Custom Reports dialog box and click Edit, the Task Report dialog box for the Critical Tasks will appear (Fig. 6.23). Most of Microsoft® Project's editing dialog boxes have three tabs:

▪ Definition
▪ Details
▪ Sort

Definition

The Definition tab is the default tab. You will use the fields in this tab to select and/or filter the information you want to include in your report. The Period field allows you to specify time intervals for your report. The options are as follows:

▪ Entire Project
▪ Years
▪ Half years
▪ Quarters
▪ Months
▪ Thirds of months
▪ Weeks
▪ Days

The Table field allows you to specify which table fields you want to display in your report. The Entry table is the default. The other Table field options are as follows:

▪ Baseline
▪ Constraint Dates
▪ Cost
▪ Delay
▪ Earned Value
▪ Earned Value Cost Indicators
▪ Earned Value Schedule Indicators

Resource Usage as of Wed 6/28/06
Chapter3a

	8/14/05	8/21/05	8/28/05	9/4/05	9/11/05	9/18/05	9/25/05	10/2/05	10/9/05	Total
Carpenter - Cl1	8 hrs	64 hrs	88 hrs	160 hrs	72 hrs	104 hrs	56 hrs		16 hrs	568 hrs
Clear Site	8 hrs	8 hrs								16 hrs
Building Layout		16 hrs								16 hrs
Form Slab		32 hrs								32 hrs
Prepare Slab for Pour		8 hrs	16 hrs							24 hrs
Rough Frame Walls			48 hrs	72 hrs						120 hrs
Rough Frame Roof			24 hrs	48 hrs						72 hrs
Install Doors and Windows				32 hrs						32 hrs
Install Wall Insulation				8 hrs	24 hrs					32 hrs
Install Shingles						64 hrs				64 hrs
Exterior Finish Carpentry						24 hrs				24 hrs
Hang Drywall							8 hrs			8 hrs
Place Cabinets							48 hrs			48 hrs
Interior Finish Carpentry					16 hrs	16 hrs				32 hrs
Prepare Driveways and Walks					32 hrs					32 hrs
Punch List									16 hrs	16 hrs
Carpenter - CL2			48 hrs	120 hrs	24 hrs		8 hrs			200 hrs
Rough Frame Walls			48 hrs	72 hrs						120 hrs
Rough Frame Roof				48 hrs	24 hrs					72 hrs
Place Cabinets							8 hrs			8 hrs
Drywall Finisher						16 hrs	32 hrs			48 hrs
Finish Drywall						16 hrs	32 hrs			48 hrs
Laborer	16 hrs	94 hrs	152 hrs	176 hrs	96 hrs	104 hrs	88 hrs			726 hrs
Clear Site	16 hrs	16 hrs								32 hrs
Building Layout		30 hrs								30 hrs
Form Slab		32 hrs								32 hrs
Prepare Slab for Pour		16 hrs	32 hrs							48 hrs
Pour Slab			24 hrs							24 hrs
Rough Frame Walls			48 hrs	72 hrs						120 hrs
Rough Frame Roof			48 hrs	24 hrs						72 hrs
Install Doors and Windows				16 hrs						16 hrs
Install Wall Insulation				48 hrs						48 hrs
Install Shingles				16 hrs	32 hrs					48 hrs
Exterior Finish Carpentry					16 hrs	16 hrs				32 hrs
Hang Drywall					48 hrs					48 hrs
Place Exterior Brick							64 hrs			64 hrs
Interior Finish Carpentry						24 hrs				24 hrs
Prepare Driveways and Walks						64 hrs				64 hrs
Pour Driveways and Walks							24 hrs			24 hrs
Mason							64 hrs			64 hrs
Place Exterior Brick							64 hrs			64 hrs
Mixer							32 hrs			32 hrs
Place Exterior Brick							32 hrs			32 hrs

Figure 6.21 Resource Usage Report

Figure 6.22 Custom Reports Dialog Box

- Entry
- Export
- Hyperlink
- Rollup Table
- Schedule
- Summary
- Tracking
- Usage
- Variance
- Work

The Filter field allows you to narrow the list of tasks and select a group of tasks to display. The Filter field options are as follows:

- All Tasks
- Completed Tasks
- Confirmed
- Cost Greater Than...
- Cost Overbudget
- Created After
- Critical
- Date Range...
- In Progress Tasks
- Incomplete Tasks
- Late/Overbudget Task Assigned To...
- Linked Fields

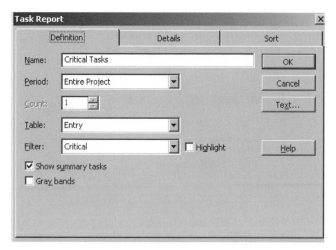

Figure 6.23 Task Report Dialog Box

- Milestones
- Resource Group
- Should Start By...
- Should Start/Finish By
- Slipped/Late Progress
- Slipping Tasks
- Summary Tasks
- Task Range...
- Tasks With A Task Calendar Assigned
- Tasks With Attachments
- Tasks With Deadlines
- Tasks With Estimated Durations
- Tasks With Fixed Dates
- Task/Assignments With Overtime
- Top Level Tasks
- Unstarted Tasks
- Using Resource in Date Range
- Using Resource...
- Work Overbudget

If you click the Text button, the Text Styles dialog box will open (Figure 6.24). Use the Text Styles dialog box is to specify the font characteristics for your report.

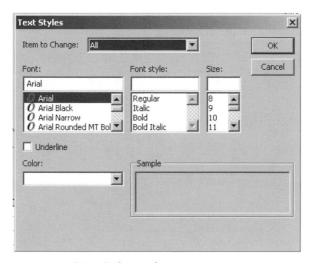

Figure 6.24 Text Styles Dialog Box

Details

This tab allows you to specify the elements that you want to appear on your report printout (Fig. 6.25) such as Notes, Objects, Predecessors, Schedule, Cost, and Work.

Sort

This tab allows you to change the task sort criteria. There are three levels of sort specifications (Fig. 6.26).

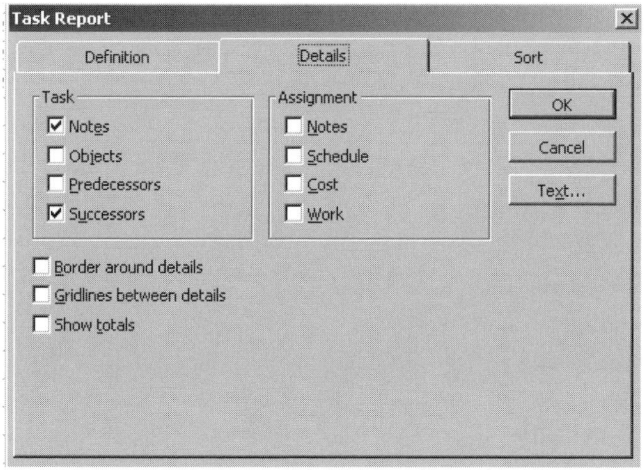

Figure 6.25 Details Tab Option

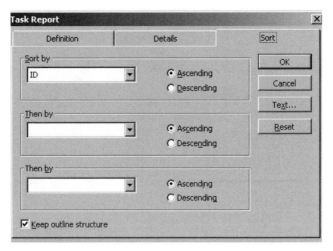

Figure 6.26 Sort Tab Option

The Sort by field options are as follows:

- %—Complete, Work Complete
- Actual—Cost, Duration, Finish, Overtime Cost, Overtime Work, Start, Work
- ACWP
- Assignment—Delay, Units
- Baseline – 1 to 10 —Costs, Duration, Finish, Start, Work
- BCWP
- BCWS
- Constraint—Date, Type
- Contact
- Cost—1 to 10, Rate Table, Variance
- CPI
- Created
- Critical
- CV, CV%
- Date—1 to 10
- Deadline
- Duration—1 to 10, Variance
- EAC
- Early Finish
- Early Start
- Earned Value Method

- Effort Driven
- Estimated
- External Task
- Finish—1 to 10, Slack, Variance
- Fixed Cost—Accrual
- Flag—1 to 20
- Free Slack
- Hide Bar
- Hyperlink—Address, Href, SubAddress
- ID
- Ignore Resource Calendar
- Late—Finish, Start
- Level Assignments
- Leveling—Can Split, Delay
- Linked Fields
- Marked
- Milestone
- Name
- Notes
- Number—1 to 20
- Objects
- Outline Code—1 to 20, Level, Number
- Overallocated
- Overtime—Cost, Work
- Physical % complete
- Predecessors
- Preleveled—Finish, Start
- Priority
- Project
- Recurring
- Regular Work
- Remaining—Cost, Duration, Overtime Cost, Overtime Work, Work
- Resource—Group, Initials, Names, Phonetics, Type
- Resume
- Rollup
- SPI
- Start—1 to 10, Slack, Variance

- Status
- Stop
- Subproject—File, Read Only
- Successors
- Summary
- SV
- Task Calendar
- TCPI
- Text—1 to 30
- Total Slack
- Type
- Unique—ID, ID Predecessor, ID Successor
- VAC
- WBS
- Work—Contour, Variance

The most common sort fields are Early Start and Critical. After you've made your first Sort by selection, you can further filter the information in your report by entering information into the Then by fields.

Copying Reports

The Copy button is only available from the Custom Reports dialog box. After you print a report, you can later make copies of it. If you don't specify a new name for the copied report, Microsoft® Project will automatically insert "Copy of" in front of the existing filename. For example, if you make a copy of the Critical Tasks report and don't rename it, the report "Copy of Critical Tasks" will be displayed in the Custom Reports list (Figure 6.27).

Figure 6.27 Copy of Critical Tasks

PRACTICE THIS SKILL

1. Open Chapter6.mpp on the CD.
2. Click Reports on the View menu.
3. Click Custom and then click Select.
4. In the Reports list, click the report you want to copy, and click Copy.
5. Type "Practice Report" in the Name box.
6. Click OK.
7. Your new report should now appear in the Custom Reports list.

Index

ABOUT THE NATIONAL ASSOCIATION OF HOME BUILDERS

The National Association of Home Builders is a Washington-based trade association representing more than 225,000 members involved in home building, remodeling, multifamily construction, property management, trade contracting, design, housing finance, building product manufacturing, and other aspects of residential and light commercial construction. Known as "the voice of the housing industry," NAHB is affiliated with more than 800 state and local home builders associations around the country. NAHB's builder members construct about 80 percent of all new residential units, supporting one of the largest engines of economic growth in the country: housing.

 Join the National Association of Home Builders by joining your local home builders association. Visit www.nahb.org/join or call 800-368-5242, x0, for information on state and local associations near you. Great member benefits include:

- Access to the **National Housing Resource Center** and its collection of electronic databases, books, journals, videos, and CDs. Call 800-368-5254, x8296 or e-mail nhrc@nahb.org
- **Nation's Building News**, the weekly e-newsletter containing industry news. Visit www.nahb.org/nbn
- **Extended access to www.nahb.org** when members log in. Visit www.nahb.org/login
- **Business Management Tools** for members only that are designed to help you improve strategic planning, time management, information technology, customer service, and other ways to increase profits through effective business management. Visit www.nahb.org/biztools
- **Council membership**:
  Building Systems Council: www.nahb.org/buildingsystems
  Commercial Builders Council: www.nahb.org/commercial
  Building Systems Council's Concrete Home Building Council: www.nahb.org/concrete
  Multifamily Council: www.nahb.org/multifamily
  National Sales & Marketing Council: www.nahb.org/nsmc
  Remodelors™ Council: www.nahb.org/remodelors
  Women's Council: www.nahb.org/womens
  50+ Housing Council: www.nahb.org/50plus

 BuilderBooks, the book publishing arm of NAHB, publishes inspirational and educational products for the housing industry and offers a variety of books, software, brochures, and more in English and Spanish. Visit www.BuilderBooks.com or call 800-223-2665. NAHB members save at least 10% on every book.

 BuilderBooks Digital Delivery offers over 30 publications, forms, contracts, and checklists that are instantly delivered in electronic format to your desktop. Visit www.BuilderBooks.com and click on Digital Delivery.

 The **Member Advantage Program** offers NAHB members discounts on products and services such as computers, automobiles, payroll services, and much more. Keep more of your hard-earned revenue by cashing in on the savings today. Visit www.nahb.org/ma for a comprehensive overview of all available programs.